MoCap for Artists

MoCap for Artists

Workflow and Techniques for Motion Capture

Midori Kitagawa and Brian Windsor

Focal Press
Taylor & Francis Group

NEW YORK AND LONDON

First published 2008

This edition published 2015 by Focal Press
70 Blanchard Road, Suite 402, Burlington, MA 01803

and by Focal Press
2 Park Square, Milton Park, Abingdon, Oxon OX14 4RN

Focal Press is an imprint of the Taylor & Francis Group, an informa business

Notices
Practitioners and researchers must always rely on their own experience and knowledge in evaluating and using any information, methods, compounds, or experiments described herein. In using such information or methods they should be mindful of their own safety and the safety of others, including parties for whom they have a professional responsibility.

Product or corporate names may be trademarks or registered trademarks, and are used only for identification and explanation without intent to infringe.

Library of Congress Cataloging-in-Publication Data

Kitagawa, Midori.
  MoCap for artists : workflow and techniques for motion capture / by Midori Kitagawa and Brian Windsor.
      p. cm.
  Includes index.
  ISBN-13: 978-0-240-81000-3 (pbk. : alk. paper) 1. Computer animation. 2. Motion—Computer simulation.
3. Three-dimensional imaging. I. Windsor, Brian. II. Title.

  TR897.7.K58 2008
  006.6'96—dc22   2008000453

British Library Cataloguing-in-Publication Data
A catalogue record for this book is available from the British Library.

ISBN: 978-0-240-81000-3 **(pbk)**

Cover Designer: Alisa Andreola
Cover Direction: Alisa Andreola
Cover Image: Eddie Smith and Patrick Dunnigan

Typeset by Charon Tec Ltd (A Macmillan Company), Chennai, India

Contents

Supplementary Resources Disclaimer

Additional resources were previously made available for this title on CD. However, as CD has become a less accessible format, all resources have been moved to a more convenient online dow nload optio n.

You can find these resources available here: www.routledge.com/ 9780240 810003

Please note: Where this title mentions the associated disc, please use the downloadable resources instead.

Acknowledgments

We would like to thank the following people who have helped us put this book together.

Eddie Smith and Patrick Dunnigan for a still image from their animation "Motion Captive" for the front cover, the 3D hand model in Figures 6.8 and 6.9, and allowing us to include "Motion Captive" on this book's CD, Frank Dufour for designing sound for "Motion Captive," Singkham Khamnouane for the 3D character models in Figures 6.2 and 10.5a, b, Patrick Kenney for the 3D hand model for Chapter 9, David Hanson and Hanson Robotics for the 3D head model in Figure 10.9, Jason Huang for the 3D plane in Figures 12.1 and 12.4, Dane Sigua for the 3D character model in Figures 12.4, 12.5a–c, and 12.6a, b, Ken Murano and Bill Lorton for general fact checking, Fran Kalal, Tim Daoust, Mike Maloy, Josh Huber, Eric Camper, and Brent Haley for help with motion capture and mocap images, Jeff Senita, Jason Huang, and Tony Tyler for their assistance at the UTD mocap studio, Thomas Linehan, Director of the Arts and Technology Program at UTD, for his support, Bruce Barnes for his assistance in literature research and proofreading, and our families for their love and support.

Introduction

If anyone asks us "Do you think motion capture will replace key-frame animation?" we will say "No."

Motion capture (mocap) is a very fast and accurate way to bring human motion into a 3D computer animation, but it is not always the best way. Mocap technology exhibits its remarkable strengths for some projects while other methods, such as key-framing, work much better for some other projects. We all need to make sure that whichever the method we decide to use, that's the most effective method for the project.

In this book, we try to give you a basic overview of motion capture based off of the most popular method which is currently (circa 2007) optical motion capture that tracks retro-reflective markers. We don't even try to say that our way is the only way or always the best way, just that it works for us. You will find many different ways to tackle problems as you go along and find different ways to think about mocap data. Every person we know in the mocap industry has a different idea about how to do things and very good reasons for doing them that way. It's a big pond and there's room for lots of different ideas and methods. If you use motion capture you'll create some of your own methods.

Our goal is to help those who are involved in 3D computer animation and games have a better understanding of motion capture as a whole so that they can decide if they need it for a project, and if they do, how it can be used. One of the things we try to do in this book is to have a mix of showing you how to do things and telling you the theory behind it. Neither one of us was interested in writing a "manual" nor a definite guide that tells you exactly what you need to do with specific mocap hardware and software. We wanted to make the information as broad as possible and applicable to as many situations as possible. So there will be times you need to rely on the user manuals from your hardware and software providers.

We are currently using optical motion capture systems at our workplaces and our pipelines involve Maya and MotionBuilder, where MotionBuilder is currently the only widely available motion editing tool with adequate functionality. Many examples in this book are products of the pipelines. However, let us emphasize this again: this is not the only way. Other applications can work in your pipeline.

Photography did not replace painting as some predicted. It has found its own place in visual arts. We believe that motion capture is finding its own place in motion picture and interactive arts.

We hope this book will help you get started with mocap and give you a roadmap to how it all works. We also hope to see results of your creative endeavors.

1 An Overview and History of Motion Capture

1.1 About This Book

Motion capture (mocap) is sampling and recording motion of humans, animals, and inanimate objects as 3D data. The data can be used to study motion or to give an illusion of life to 3D computer models. Since most mocap applications today require special equipment there are still a limited number of companies, schools, and organizations that are utilizing mocap technology. Outside the film industry, army, and medicine, there are not too many people who know what mocap is. However, most people, even small children, have seen the films, games, and TV commercials for which mocap technology is used. In that sense mocap is in our everyday life.

The main goal of this book is to help you understand steps and techniques that are essential in a workflow or pipeline for producing a 3D character animation using mocap technology. Capturing data using mocap equipment is, of course, the essential part of the pipeline, but equally important are the things we do before and after capturing data, that is preproduction (planning), data cleaning and editing, and data applications. Without a well-thought-out preproduction for a project, the project is destined to fail or go through preventable difficulties. It cannot be emphasized enough that good preproduction is a key to the success of a project. After capture sessions, data needs to be cleaned, edited, and applied to a 3D model. Applications are getting better every year but they are tools, that is, technology does not create arts, you do. You are the creator and decision-maker.

Another key to success is setting up a reliable pipeline that suits your needs and environment. We've heard about production companies deciding to use mocap for particular projects, believing that mocap would cut their production cost and time, and giving up on mocap quickly after finding that mocap was neither quick nor cheap. Mocap disaster stories are often caused by the lack of a reliable production pipeline. Mocap technology can be effectively used once a production pipeline is established. For the first project or two, you will be hammering kinks out of your production pipeline while the project is moving through the pipeline. Thus, greater productivity shouldn't be expected immediately after introducing mocap technology into the production environment.

This book is written for artists, educators, and students who want to create 3D animation for films and games using mocap technology. Familiarity with basic concepts of 3D animation, such as the principles of animation and inverse kinematics is expected. In the rest of this chapter we look at the history of mocap and types of mocap systems. We detail the preproduction in Chapter 2 and pipeline in Chapter 3, and introduce you to cleaning and editing data in Chapter 4. Skeletal data editing is explained in Chapter 5. Chapters 6–8 are about applying data to 3D models. In Chapter 6 we show you simple cases of data applications. In Chapter 7' we discuss mapping multiple motions and taking motions apart. In Chapter 8' we explain how you can integrate data into rigs. Special issues about hand capture are discussed in Chapter 9, facial capture in Chapter 10, and puppetry capture in Chapter 11. Chapter 12 covers mocap data types and formats, and mathematical concepts that are useful to know when you are setting up or troubleshooting a production pipeline.

We suggest that you read through this book once before you start a mocap project, and read it again as you go through your project pipeline.

1.2 History of Mocap

The development of modern day mocap technology has been led by the medical science, army, and computer generated imagery (CGI) field where it is used for a wide variety of purposes. It seems that mocap technology could not exist without the computer. However, there were early successful attempts to capture motion long before the computer technology became available. The purpose of this section is to shed light on some of the pioneers in mocap in the 19th and 20th centuries: this is not our attempt to list all the achievements on which today's mocap technology is built upon.

1.2.1 Early attempts

Eadweard Muybridge (1830–1904) was born in England and became a popular landscape photographer in San Francisco. It is said that in 1872 Leland Stanford (California governor, president of the Central Pacific Railroad, and founder of Stanford University) hired Muybridge to settle a $25,000 bet on whether all four feet of a horse leave the ground simultaneously or not. Six years later Muybridge proved that in fact all four feet of a trotting horse simultaneously get off the ground. He did so by capturing a horse's movement in a sequence of photographs taken with a set of one dozen cameras triggered by the horse's feet.

Muybridge invented the **zoopraxiscope**, which projects sequential images on disks in rapid succession, in 1879. The zoopraxiscope is considered to be one of the earliest motion picture devices. Muybridge perfected his technology for sequential photographs and published his photographs of athletes, children, himself, and animals. His books, *Animals in Motion* (1899) and *The Human Figures in Motion* (1901), are still used by many artists, such as animators, cartoonists, illustrators, and painters, as valuable references. Muybridge, who had a colorful career and bitter personal life, is certainly a pioneer of mocap and motion pictures (Figure 1.1).

Born in France, in the same year as Muybridge, was **Etienne-Jules Marey**. Marey was a physiologist and the inventor of a portable **sphygmograph**, an instrument that records the pulse and blood pressure graphically. Modified versions of his instrument are still used today.

In 1882 Marey met Muybridge in Paris and in the following year, inspired by Muybridge's work, he invented the chronophotographic gun to record animal locomotion but quickly abandoned it. In the same year he invented a chronophotographic fixed-plate camera with a timed shutter that allowed him to expose multiple images (sequential images of a movement) on a plate. The camera

Figure 1.1 Mahomet Running, *Eadweard Muybridge, 1879*

initially captured images on glass plates but later he replaced glass plates with paper film, introducing the use of film strips into motion picture. The photographs of Marey's subject wearing his mocap suit show a striking resemblance to skeletal mocap data (Figures 1.2 and 1.3).

Marey's research subjects included cardiology, experimental physiology, instruments in physiology, and locomotion of humans, animals, birds, and insects. To capture motion, Marey used one camera while Muybridge used multiple cameras. Both men died in 1904, leaving their legacies in arts and sciences.

In the year after Muybridge and Marey passed away **Harold Edgerton** was born in Nebraska. Edgerton developed his photographic skills in the early 1920s while he was a student at University of Nebraska. In 1926 while working on his master's degree in electrical engineering at the Massachusetts Institute of Technology (MIT), he realized that he could observe the rotating part of a motor as if the motor were turned off by matching the frequency of the strobe's flashes to the speed of the motor's rotation. In 1931 Edgerton developed the stroboscope to freeze fast moving objects and capture them on film. Edgerton became a pioneer in high-speed photography (Figures 1.4 and 1.5).

Edgerton designed the first successful underwater camera in 1937 and made many trips aboard the research vessel *Calypso* with French oceanographer Jacques Cousteau. He designed and built deep

Figure 1.2 *Etienne-Jules Marey's mocap suit, 1884*

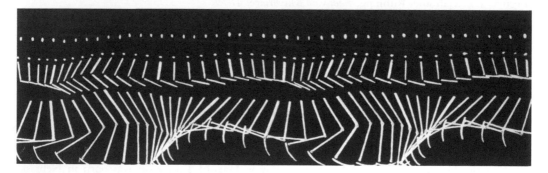

Figure 1.3 *Motion photographed by Etienne-Jules Marey, 1886*

sea electronic flash equipment in 1954. Edgerton ended his long career as an educator and researcher at MIT when he passed away in 1990.

1.2.2 Rotoscoping

Max Fleischer, born in Vienna in 1883, moved to the U.S. with his family in 1887. When he was an art editor for *Popular Science Monthly*, he came up with an idea for producing animation by tracing live

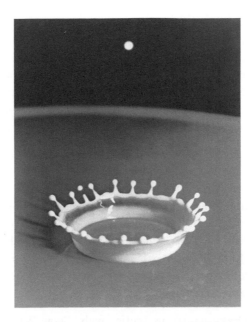

Figure 1.4 Milk-Drop Coronet, *Harold Edgerton, 1957*

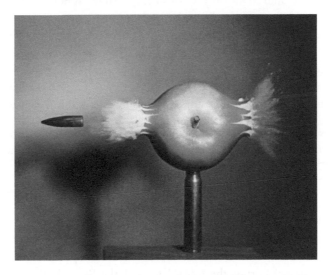

Figure 1.5 Shooting the Apple, *Harold Edgerton, 1964*

action film frame by frame. In 1915 Fleischer filmed his brother, David, in a clown costume and they spent almost a year making their first animation using **rotoscope**. Fleischer obtained a patent for rotoscope in 1917. World War I ended in 1918 and in the following year he produced the first animation in the "Out of the Inkwell" series and he also established Out of the Inkwell, Inc., which was later renamed as Fleischer Studio. In the "Out of the Inkwell" series, animation and live action were cleverly mixed and Fleischer himself interacted with the animation characters, Koko the Clown and Fitz the dog. In 1924, 4 years before Disney's "Steamboat Willie," Fleischer produced the first animation with a synchronized soundtrack. Fleischer Studio animated characters from the comics, such as Popeye the Sailor and Superman. Betty Boop first appeared in Fleischer's animation and later became a comic strip character. Fleischer's early 30s animations were filled with sexual humor, ethnic jokes, and gags. When the Hays Production Code (censorship) laws became effective in 1934 it affected Fleischer Studio more than other studios. As the result, Betty Boop lost her garters and sex appeal.

In 1937, after almost 4 years of production, **Walt Disney** (1901–1966) presented the first feature length animation, "Snow White and Seven Dwarfs." "Snow White" was enormously successful. Paramount, the distributor of Fleischer's animations, pressured Max and David Fleischer to produce feature length animations. They borrowed money from Paramount and produced two features, "Gulliver's Travels" (1939) and "Mr. Bug Goes to Town" (1941). Neither film did well at the box office. After the failure of "Mr. Bug," Paramount fired the Fleischer brothers and changed the studio's name to Famous Studios. In the 1950s Max Fleischer sued Paramount over the distribution of his animations. Before his death in 1972, he signed a Betty Boop merchandising deal for King Features, a unit of the Hearst Corporation.

Rotoscoping was used in Disney animations, starting with "Snow White." Later Disney animations characters were highly stylized and rotoscoping became a method for studying human and animal motions. Disney's classic animations on DVDs, such as "Snow White" and "Bambi," contain live action film footages from the Disney archive. Comparison between film footages and the corresponding scenes in the animations reveals skillful and selective use of rotoscoping by Disney animators. They went above and beyond rotoscoping. The success of "Snow White" can be attributed to Walt Disney's detailed attention to the plot, character development, and artistry. There are some beautiful scenes in the Fleischers' feature length animations. However, they do not have plots that can sustain the audience's interests until the end of the movie nor characters that make the audience care about them.

Both Max Fleischer and Walt Disney were highly innovative individuals; however, it is sadly true that "Disney's memory belongs to the public; Max's to those who remember him by choice" (Heraldson, 1975).

1.2.3 Beginning of digital mocap

Research and development of digital mocap technology started in pursuit of medical and military applications in the 1970s. The CGI industry discovered the technology's potentials in the 1980s. Since some of this book's readers weren't born in the 1980s, let's recall the 1980s. In the 1980s there were floppy disks that were actually floppy and most computers were equipped with monochrome monitors; some with calligraphic displays. To view color images, for example rendered animation frames, images had to be sent to a "frame buffer," which was often shared by multiple users due to its cost. Large computers were housed in ice cold server rooms. The noise of dot matrix printers filled offices. Ray-tracing and radiocity algorithms were published in the 1980s. Renderers based on these algorithms required a supercomputer or workstations to render animation frames in a reasonable amount of time. Personal computers weren't powerful enough. (Ray-tracing and radiocity didn't become widely available until the computing power improved.) CPUs, memories, storage devices, and applications were more expensive than today. Wavefront Technologies developed and marketed the first commercial off-the-shelf 3D computer animation software in 1985. Only a handful of computer animation production companies existed. Most of the animations that they produced were "flying logos" for TV commercials or TV programme's opening sequences. These were often 15 to 30 seconds long per piece. The readers who saw "Brilliance" (also called "Sexy Robot") in the 1980s probably still remember the astonishment of seeing a computer generated character, a shiny female robot, moving like a real human being.

"Brilliance" was produced by Robert Abel and Associates for the National Canned Food Information Council and was aired during the 1985 Super Bowl. They invented their own method for capturing motion for the project. They painted black dots on 18 joints of a female model and photographed her action on a swivel stool from multiple angles. The images were imported into Silicon Graphics workstations and a number of applications were employed to extract the information necessary to animate the CGI robot. They didn't have enough computing power to render frames for the 30 second piece in house. So, in the final 2 weeks before the project deadline they borrowed VAX 11/750 computers around the country to render. The final product was a ground breaking piece and is regarded as a milestone in the history of CGI.

While "Brilliance" was the first successful application of mocap technology in CGI, "**Total Recall**" was the first failed attempt to use mocap in a feature film. Metrolight Studios was one of the production companies contracted to produce effects for the 1990 science fiction film starring Arnold Schwarzenegger and Sharon Stone. Metrolight decided to use mocap to create an animation sequence of moving skeletons for the scene in which Schwarzenegger's character goes through a large airport security X-ray machine, along with other people and a dog. (Still images from the scene popped up in news media after the 9/11 tragedy regarding security vs. privacy issues.) An operator of an optical mocap equipment company was sent to a location with a mocap system. A Metrolight team followed the operator's instruction while capturing performances by Schwarzenegger and other performers. They went home believing that the capture session had gone well and the mocap company would deliver the captured data after processing and cleaning. However, Metrolight never received usable data and had to give up using mocap for the scene.

It is not certain if the operator did not know his company's equipment well and made critical mistakes or the system was faulty. One of the lessons that we can learn from Metrolight's unfortunate experience is that if you need to rely on an external expert in capture sessions, make sure that you hire a service provider with a good track record and references. Another is to check the data after capturing the range of motion and one or two shots. You don't want your capture subjects to wait for a long time. Don't process all the captured data during capture sessions, but check what kind of data quality you are getting, especially at a location site that you are not familiar with. If you need to recalibrate, recalibrate the system. Recalibrate after lunch or a long break and again before you wrap up sessions. Also be prepared to have a follow-up shoot in case the director, clients, you, or any other decision-makers make changes after the first shoot, or data from the first shoot has problems. Mocap wasn't used as planned but "Total Recall" won an Academy Award (Special Achievement Award) for its special effects. Mocap technology had to wait for a few more years to come into the limelight.

Released in 1995, **FX Fighter** is the first real-time fighting game with 3D characters in 3D environments. It is also one of the first video games that used mocap technology to give realism to 3D characters' movements. Game characters are animated in real time by the user input using a set of motion captured actions, for example, running, walking, and kicking. Pieces of actions are played in such a way that the player does not notice the transition from one action to another giving an impression that the player is fully in control of a game character's movement. The game's success encouraged other game companies to use mocap in their games.

Since these pioneering efforts in the 1980s and 1990s we have seen remarkable development and achievement in digital mocap. In recent years, in addition to medicine, military, and entertainment, mocap applications have been found in many other fields. Various sports use mocap to analyze and enhance athletes' performances and prevent injuries. Designers use mocap to understand users' movements, constraints, and interactions with environments and to design better products. Engineers use mocap to analyze human movements and design robots that walk like us. Art historians and educators use mocap to archive and study performances by dancers and actors. For instance, in 1991 an intricate performance by legendary French mime Marcel Marceau (1923–2007) was captured at the Ohio State University to preserve his arts for future generations.

1.3 Types of Mocap

Mocap systems commercially available today can be categorized into three main groups: optical systems, magnetic systems, and mechanical systems. Each type has strengths and weaknesses, which we will look at in this section. We will not look at ultrasonic and inertial systems because they are very infrequently used in the entertainment environment.

1.3.1 Optical mocap systems

Most optical mocap systems are primarily designed for medical applications. The first commercially available optical system developed with CGI applications in mind was the Vicon 8 system. A typical optical system consists of 4 to 32 cameras and a computer that controls the cameras. With most optical systems capture subjects wear markers, where markers are either reflective (passive) or light emitting (active). Passive markers are made of reflective materials and their shapes are spherical, semi-spherical, or circular. Shapes and sizes of markers depend on the camera resolutions and capture subjects (e.g., smaller markers are used for facial and hand captures). Passive markers are attached directly to a capture subject's skin or Velcroed to a mocap suit, which is a full-body unitard made of stretchy materials, such as spandex. Cameras in a passive marker system are equipped with light-emitting diodes (LEDs) and the lights emitted by the LEDs are reflected by markers (Figure 1.6). On the other hand, markers in an active marker system are LEDs. Some active marker systems illuminate one LED at a time, eliminating the need for identifying each marker. Others illuminate all LEDs at once. Modulating the amplitude or frequency of each LED allows such systems to identify markers. Some of the latest active marker systems work in natural lighting conditions, that is, they can capture subjects in various costumes at locations outside studios; however, lighting must be carefully controlled for most optical systems, especially passive marker systems.

Figure 1.6 *Vicon camera with LEDs*

Cameras in an optical system capture the lights reflected or emitted by markers at speeds somewhere between 30 and 2000 samples per second. At least two cameras need to see a marker in order to determine the 3D position of the marker, although three or more are preferred for accuracy. Sometimes a capture subject herself/himself, another capture subject, or a prop hides (occludes) some of the markers on the subject. For instance, when a subject lies flat on the stomach, the markers on the subject's front will be occluded. When markers are occluded, no camera sees them and it results in loss of data. There are data editing techniques and tools to make up for missing data but when too many markers are occluded or the duration of an occlusion is too long, it is impossible to fix the problem. Optical data generated by a state of the art system is very clean and accurate when it does not suffer from occlusion problems.

Marker configurations are flexible with optical systems. You can use the marker configurations that the system manufacturer provides you with or you can design your own that suits your needs. A relatively large number of markers can be tracked simultaneously, for example, up to 200 markers with a 16 camera system. Since capturing multiple subjects at once tends to cause occlusion problems, capture one subject at a time if it is not crucial to capture multiple performers together. When performers interact with each other and the synchronization among them is important, capture multiple subjects simultaneously. Capture subjects can move freely in a capture volume because no equipment or wires are connected to them.

Optical systems' real-time visual feedback during capture is often limited to stick figures, although linking a mocap's real-time output to a specific real-time application such as MotionBuilder will render real-time results. Other systems such as the Giant Studios system readily render real-time characters directly in system. Recorded data is still processed to compute the trajectories of the markers in a rather extensive post-processing to get the best, most stable results. Rotational data can be computed in real time, but is usually computed from positional data in post-processing.

Among the markerless mocap technologies that recently emerged, Mova's Contour Reality Capture system is an optical system that captures the continuous skin surface of a moving capture subject, instead of a small number of points on a capture subject. A capture subject wears a phosphorescent makeup and two sets of cameras capture the texture and geometry of the subject in a movement. We will see how new markerless systems will compete with other systems.

Advantages of optical systems:

- Optical data is accurate.
- Capture rate is high.
- Multiple subjects can be captured simultaneously.
- A large number of markers can be used.
- Marker configurations can be changed easily, depending on project goals.
- Optical systems' capture subjects can move freely in a capture volume.
- Capture volume can be larger than most other systems.
- Skeletal data can be generated.

Disadvantages of optical systems:

- Extensive post-processing is required
- Rotational data needs to be computed from positional data in the post-processing.
- Markers can be occluded by capture subjects or props, resulting in loss of data.
- Lighting needs to be controlled for most optical systems, especially passive systems.
- Real-time visual feedback is limited to stick figures.
- Hardware is often more expensive than other types of mocap equipment.

1.3.2 Magnetic mocap systems

Magnetic (electromagnetic) mocap systems are sometimes called magnetic trackers. The systems were derived from the sensors placed on a military aircraft pilot's helmet to track the pilot's head position and orientation for the helmet-mounted display. With a magnetic mocap system, 12 to 20 tracking sensors are placed on a capture subject to measure spatial relationship to a magnetic transmitter. The tracking sensors output their translations and orientations. Hence, no post-processing is required to compute rotations. This fact allows magnetic systems to be used for real-time applications.

Tracking sensors are not occluded by capture subjects or props made of non-metallic materials, which is an advantage over optical systems. However, they are prone to magnetic and electrical interferences caused by metal objects and electronics in the environments. Interferences can result in distorted output. Building structures with high-conductivity metals are not suitable as capture spaces for magnetic systems. The wiring and batteries for tracking sensors may limit capture subjects' movements. Moreover, tracking sensors' batteries need to be recharged every few hours.

Magnetic systems can be divided into two groups. One group uses direct current (DC) electromagnetic fields and the other uses alternating current (AC) fields. AC systems (e.g., Liberty by Polhemus) are very sensitive to aluminum and copper. DC systems (e.g., Flock of Birds by Ascension) are sensitive to iron and steel.

Magnetic systems' sampling rates (up to 144 or 240 samples per second) are lower than optical systems and magnetic data tends to be noisy. Tracking sensors' configurations cannot be changed as freely as optical systems' marker configurations. Magnetic systems can capture multiple performers simultaneously with multiple setups. Magnetic systems' capture volumes are normally smaller than optical systems'. One of the biggest advantages of magnetic systems is their cost; magnetic systems are less expensive than optical systems.

Advantages of magnetic systems:

- Position and orientation are available without post-processing.
- Real-time feedback allows real-time applications.
- Tracking sensors are not occluded by non-metallic objects.
- Multiple performers can be captured simultaneously with multiple setups.
- Magnetic systems are less expensive than optical systems.

Disadvantages of magnetic systems:

- Tracking sensors are prone to magnetic and electrical interferences.
- Wiring and batteries for tracking sensors can limit capture subjects' movements.
- Magnetic sensors have a lower sampling rate than most optical systems.
- Magnetic data tends to be noisy.
- Tracking sensors' configurations are hard to change.
- Magnetic systems' capture volumes are normally smaller than optical systems'.

1.3.3 Mechanical mocap systems

Mechanical (exo-skeletal) mocap systems directly measure joint angles of a capture subject who wears an articulated device that consists of straight rods and potentiometers. Straight rods are linked with potentiometers at the joints of the body, designed to measure joint angles as the capture subject moves. The device looks like an exo-skeleton. Other types of mechanical systems include data gloves and digital armatures.

Mechanical systems are real time, relatively inexpensive, free of occlusion, free from magnetic or electrical interferences, and highly portable. Wireless mechanical systems provide large capture volumes. A notable disadvantage of mechanical systems is that they do not measure global translation very well. They measure it using accelerometers, but the data can still "slide" and "slip" a little. They do a poor job when the feet leave the floor. If a capture subject jumps up, the data will normally not follow the jump and the data will stay on the floor. If a character walks up stairs, the data will never go up in the air but look as if it were walking in place. Magnetic sensors are often added to mechanical systems to correct this problem. The joints in articulated exo-skeletal systems are simple hinge joints, although we, humans, have other kinds of joints, such as ball and socket joints, gliding joints, saddle joints, and pivot joints. This means that articulated exo-skeletal systems restrict capture subjects' movement at their joints. Also the device's volume and breakability restrict subjects' movement, for example, a capture subject wearing an articulated exo-skeletal system probably doesn't want to roll around on a floor since it hurts and breaks the device. ShapeWrap, developed by Measurand, uses flexible fiber-optic tapes and is more durable than rigid exo-skeletal systems.

Advantages of magnetic systems:

- Real time
- Relatively inexpensive
- No occlusion
- No magnetic or electrical interferences
- Highly portable
- Large capture range

Disadvantages of magnetic systems:

- No global translations
- Restrictions on capture subject's movement
- Breakable
- Fixed configuration of sensors
- Low sampling rate

If you are choosing a mocap system to purchase, think about your goals first and then find a system that meets your needs the best. Before investing in an expensive mocap system, it may be wise to try out some service providers. If you like what a service provider delivers, find out what kind of setup it has.

2 Preproduction

Preproduction can be looked at as one of the most important steps in motion capture. This is the part of the process that allows us to break everything apart and prepare before either going into the motion capture space or before going into the space with a client. Preproduction can be broken down into pre-capture planning (Section 2.2), preparation for the capture (Section 2.3), and designing a production pipeline (Chapter 3).

2.1 Importance of Preproduction

Why do you want to spend time on preproduction? Because a good preproduction saves a lot of production cost and time, spares you from unnecessary pain, and results in a much better end-product. If you jump into production with little or no preproduction, we assure you that you will face problems at some point that will cause waste of motion capture time, difficulties during the capture, issues in processing, and other trouble. The majority of these can be avoided by taking the time to really go through the preproduction process meticulously. We can never stress enough how much planning and working through everything in advance is the difference between a motion capture nightmare and good motion capture experiences.

What preproduction gives you is a roadmap to how you're going to organize and accomplish all of your goals. It should answer the questions, such as "Why are you capturing anything?," "What are you capturing?," "How will you capture it?," and "What will you do with the data once it's captured?" Thinking through all of this and running a few small tests will make a huge difference in how easy or hard motion capture will be.

Many people have a preconceived notion that motion capture is not only simple but very quick and that it automatically works. This, unfortunately, is not the case. Getting good data out of a motion capture system takes patience and experience. The process will not be quick until you've done it several times and are used to how it works. Motion capture never works automatically. However, with enough planning, even the first time you work all the way through a motion capture project, you will have a clear idea about how you get from one end to the other.

2.2 Pre-capture Planning

One of the most important parts of not only enjoying mocap, but getting good data out of it, is to have everything ready to go before your mocap shoot. Much of this process is done in spreadsheets, word processors, and thumbnail sketches. We'll now take a look at the areas that need to be fleshed out before you go into a shoot.

2.2.1 Script

The script is the narrative of the story that you're trying to tell. This could be anything. It could be a screenplay, an idea for a short animation, or the shots needed for a video game. The script is one of the basic building blocks of any type of animation process.

The script allows you to bring all of your ideas together and gives a certain flow to the story that you want to tell. It not only serves as a roadmap for the client, but it also serves as a roadmap for the talent. The talent will need a script as early in the process as possible. This helps them not only prepare to be in character, but to understand what will be physically required of them.

> When we refer to the client, we're referring to whomever you are capturing the motion for. This can be for a class assignment, for yourself, for another group of students, for teachers who need some motion capture, or for paying clients in a studio. Always think of what you're doing in this relationship and give your client the best motion capture experience possible. Don't cut corners or try to pass off bad data to anyone. It will hurt your reputation, and the motion capture world is a fairly small one.
>
> Because I've worked in the private sector with motion capture, I'll at times put emphasis on saving money and time, and these are important to think about no matter what situation you're in. Even though saving money may not be relevant to a student who has access to a mocap system based on a course enrollment, still think about the impact that your decisions will have on the time it takes you to complete certain tasks.

Scriptwriters are professionals who specialize in writing scripts for films, TV programmes, animations, video games, etc. It is common for a film script to be rewritten by multiple screenwriters while a project is going through the preproduction, production, and post-production phases.

Typical scripts contain the following elements for each scene:

- *Scene Heading*: Short description of when and where a scene takes place.
- *Action*: Description of any moving elements in the scene.
- *Dialog*: Conversation between characters.
- *Character Name*: Character's name that precedes his/her line in a dialog.
- *Transition*: Editing transition between shots or scenes (e.g., cut and dissolve).
- *Shot*: Name of a camera shot (e.g., close-up shot, long shot, and tracking shot).

Most scripts go through several revisions and they are eventually torn apart to create the shot lists which we will discuss in Section 2.2.3. Because of this, use a type of word processing program where revisions can be maintained and multiple people can make comments or changes. The script is also used as the basis for the storyboard.

2.2.2 Storyboard

The storyboard is a 2D visual representation of the script. A script is turned into a set of drawings and short text that accompanies each drawing. They present essential elements of character performance, timing, staging, camera (shot size, camera move, camera angle, etc.), editing (transitions between shots, etc.), and sound (dialog, narrative, sound effect, etc.). Drawings for a storyboard are usually quick thumbnail sketches that are rough but pre-visualize the motions and emotions of the script.

Storyboards have traditionally been one of the most important steps in preproductions of animations, films, TV programmes, and games. They allow content creators, directors, performers, cinematographers, clients, and anyone else who is involved in the preproduction and/or production to get a clear picture of how the final product will look.

Most storyboard artists create storyboards in a format that allows them to make modifications easily. Some pin thumbnail sketches to a cork board with pushpins; others draw on Post-it notes and stick them to a wall. Either way, they can add, delete, replace, and rearrange any parts of the storyboard while they are trying to find the best way to tell the story.

How storyboards are made is not important. You can use pencils, markers, watercolors, pastels, computer applications, and anything else. No matter what you decide to use as a tool for storyboarding, keep in mind that you are making a storyboard to get your idea across. Storyboards do not have to look like art works because they are not. The main goal of storyboarding is to firm your ideas up and communicate with a team of artists and with decision makers, such as directors and clients. A good storyboard will result in a final product that reaches the audience's heart.

2.2.3 Shot list

The shot list is a list of actions or motions that will be compiled together to create a scene. Breaking the story into shots will give you a very good idea about how complex each scene and shot will be and also about how many talent, props, and other things will be needed for each shot. This is one of the earliest and most important steps in organizing your project. It can tell you how many days a shoot will take, what types of talent will be needed on what days, if special rigging will be needed, etc.

If there is a lot of rough action and stunt work, you may want to have a longer shoot time. If it's more dramatic with less dangerous action, you may want more acting talent. If there are special rigging shots, you'll want to find a stunt coordinator who has experience in motion capture. Also try to figure out if your motion capture equipment is capable of capturing the desired motion. If there are a lot of physically demanding shots, it will take longer to capture these shots since the talent will tire quickly. Because of this, you will get fewer shots in during a day for this type of action, so take this into account when putting together your shot list.

As a term in cinematography a shot can be defined as a continuous view filmed by a single camera with no interruption, while a scene is a place or setting for an action to take place. A scene may consist of a series of shots (or a single shot) that depicts a continuous event. Thus, you may have several shots for a scene, instead of one continuous shot. You need to think about and decide what types of camera shots you want to use based on the storyboards as well as what types of editing are feasible, where editing includes both editing motion capture data to create motions that are humanly impossible or improbable and compositing in the post-production phase.

If you have one shot that's 45 seconds of continuous running, you may not have enough physical area in your motion capture space to capture a full run for that long. The person may be through your space in a few seconds. In this case, you'd need to be able to loop smaller motions together several times to create 45 seconds of running. This is one of those types of problems you may encounter and will need to try to solve before you go into the mocap studio.

When thinking about the studio, always think about the amount of space it has as well as the capabilities. Are you going to be tracking facial, finger, and full body data all at once? If so, is this possible throughout the entire space or is it just possible in a smaller area of the space that has more cameras? What are the physical limits as far as length, width, and height of the motion capture volume? If you're doing motions that are gymnastics related and possibly need a high bar, you're going to want to make sure that this type of motion is flagged ahead of time for not fitting in the normal space as well as requiring special equipment. Use your shot list as a way to think about what you have to use for capture. We will look at shot lists in more detail in Section 2.3.4.

2.2.4 Animatic

The animatic is basically a storyboard in a time-based format that has visual elements (e.g., drawings, rough animation, camera moves, and preliminary special effects) and sound elements (e.g., dialog, sound effects, and music), where both elements are timed and edited together to fit each other. Animatics let us see how timing and cinematographic elements of the camera are working. That makes it easier for you to determine if your story is being told effectively or not. The progression reel is similar to animatics but shots and scenes in a progression reel are repeatedly replaced with animated scenes and elements that are closer to the final ones as the production proceeds.

Timing is essential, not only in telling a good story, but also in conserving money. If there are shots that can be reduced from 2 minutes down to 30 seconds, this is important and can usually be seen in an animatic. A common mistake is to try to capture really long motions or complex motions just because it's possible and not because they add to the story. Don't capture any scenes that do not contribute to the story telling. An animatic will help to see if your pacing is working and if all of your scenes are really adding to your overall story.

More and more, video game engines are being used to create animatics. Existing motion capture data can stand in as placeholder motion to get an idea of camera placement as well as timing of the shots. This is a very helpful tool since the real-time rendering capability of video games allows for instant updating and changing without long render times that you'd have with a traditional animatic rendered by hand or with animation software. A good library of generic motions for this purpose is never a bad idea.

Students often animate camera positions in their 3D computer animation, not because a camera flying through a scene adds something important to story telling, but because it is easy and fun to move a virtual camera around. If you want to animate camera moves to create an effective dynamic shot, study popular types of camera moves that are possible with a real physical camera, such as dolly, crane, and boom.

Knowledge in cinematography is essential in making good films, videos, video games, and animations, including mocap data driven ones. If you have never produced a narrative video piece, try it. Taking a video production course and a film history course is beneficial as well. There are a lot of good books on cinematography and film history too. Study and get familiar with common terms and concepts in cinematography that you need when you create storyboards and animatics.

For those who are studying animation, we highly recommend *Prepare to Board!: Creating Story and Characters for Animated Features and Shorts* by Nancy Beiman, published by Focal Press, 2007. The author explains a number of concepts and techniques that are crucial in preproduction.

2.3 Preparation for Capture

Preproduction is one area where there are many things happening at once and if any one aspect is neglected it will have a negative impact on the overall process. There are too many times when a small oversight in this area has caused either bad data or people to dislike the process so much that they'll never agree to use mocap again. We want the results to be good and the overall experience to be well thought out and enjoyable.

2.3.1 Talent

Make sure to use the proper talent. It can't be stressed enough that if you want to motion capture dancing, then use dancers, not your friend who used to dance when she was 10 but hasn't danced in years. The results are evident and the poise and confidence of the right talent come across in the data.

In one of the shoots I was involved in years ago, we had talent miscast. It was beyond the motion capture department's ability to change the talent since we were told who the talent would be by a senior member of the company. We had to say yes and make the best out of the situation.

The character was to be a very tough, aggressive leading female. The talent we got was a middle-aged dancer. What we really needed was an actor that would fit the role of the character. The talent was a wonderful person, but her motion characteristics did not match the character. If the character had been a dancer, she would have been great talent, but that wasn't the case. The motion revealed the fact that she was not right for that particular role no matter how talented she was in her field. Remember that the final look and feel of motion capture has a lot to do with the talent, and you need to get the right talent for the right situation. The right talent will produce good motion capture.

There is also something to be said for capturing people that have distinguishable physical features. One of the stunt coordinators I enjoy working with has a few definable motion traits. One of those characteristics comes with the fact that he broke his arm close to the elbow and that arm never straightens all the way out. If we use his motion for a main character, that's fine, but if we used his motion as the motion for several background characters, they all would have a slightly bent elbow. It would be fairly noticeable when half of the extras all have the same physical feature. I've heard people say "Well, you can edit that out. …" Yes, you can, but it takes much less time to mocap another person than trying to make the person look like someone else by massaging the data.

When casting for facial animation, which may or may not be done separately, do not cast based on the voice alone. Cast for the look of the character and the voice. Some people are not good for facial capture. I personally am not a great candidate for facial capture. My brow goes far enough forward and my upper eyelids come up enough that I lose any markers on my upper eyelids. I also have what would be termed chubby cheeks that shake when I laugh. If the character is supposed to have a very lean face, you don't want the motion capture data to possibly inject some secondary motion that's not wanted in your character.

If you need to capture older people or kids, capture them. Do not capture people acting like older people or younger people unless there is an extremely good overriding reason to do so. Kids, adults, and seniors all move differently, so try to cast the right talent to the age as well.

2.3.2 Marker sets

The marker set defines how many markers will be attached to your actor, prop, and anything else that will be captured and where the markers will be placed. A few of the things to think about when attaching markers are: what the limitations of your system are, how close markers can get to each other, what type of motion you are capturing, and if the markers are in logical locations.

2.3.2.1 What are the system limitations?

There are times when your system will dictate what size marker you can use and where it can be located with relationship to other markers. Because camera-based systems are reconstructing 3D data from a set of 2D images, it's important to realize that if markers are very close, the system may see two markers as one marker. Check your system's setup and marker specifications. A general rule of thumb is that once you place markers, you should still be able to place two markers in the space between any pair of adjacent markers (although you would not do that). This can sometimes be difficult to do when working on markers for facial capture. If a large amount of facial data is needed, that may necessitate a dense set of markers and that makes it very hard to space markers out properly.

2.3.2.2 What kind of motion will be captured?

The types of motion to be captured may cause you to move your markers away from the standard marker positions that you start with. For example, if your actor is constantly lying on the stomach, you may want to put more markers on the back and get rid of the ones on the front of the chest. Think if there are several markers that will be rarely or never seen because of a certain set of motions, other people, or props. If a number of markers will not be visible in a significant number of motions (maybe more than a quarter of your captures for a day), then you should make an alternative marker set. If it's only for one or two shots, it may not be worth interrupting the flow of the capture to create a different marker set. This is a judgment call you'll have to make.

We normally have to apply different marker sets when using flying harnesses as well because the harnesses' pick points and wires may interfere with markers. Since the pick points are normally on the sides of the hips the markers in those locations can be close enough to be sheared off by the wires. It's best to work around this type of setup with an alternate marker set since you'll get bad or unusable data almost every time if you use the normal marker set.

If you are doing a lot of work with physical rigging, physically demanding shots, or shots that require a lot of setup in between takes, you can always change marker sets during this time. Using your downtime to your advantage is an important area of motion capture. There always seems to be something to do, even if it's just starting to create 3D data on a different computer while other people are changing a set. It's important to maximize your time.

2.3.2.3 Know the anatomy

It's very important to understand the anatomy of whatever you're capturing. There are several anatomy books out there, but the really important issues are musculo-skeletal anatomy. We also like to refer to *Anatomy of Movement* by Blandine Calais-Germain, published by Eastland Press. When dealing with human motion, it gives a very clear view of how the body moves with a number of illustrations and it's thorough but not overly technical.

The importance of anatomy in regards to optical markers is that you are trying to represent the underlying human skeletal structure using a set of markers. You need an understanding of how bones relate to each other or what parts of the body move more than others. To decide where to attach a marker on a knee, it's important to know how to find a good location for the marker as well as be able to replicate it on the other knee without just guessing. You want to avoid areas around the joints that move a lot, and try to find locations where the bone is near the surface. For markers in between the joints, you'll have to compromise and use some locations that will move a lot. The leg is a good example where you need to place a marker on the upper leg, and the only place to put it in between the hip and knee is all muscle which will expand as the muscles are moved. You will have to put the marker someplace that will have a considerable amount of movement.

It also helps if you know how to find the bony landmarks that define the hips and other joints of the body. If you go too far below the landmarks for the hips, the motion of your marker will be heavily influenced by the upper leg; if you go too far above, it will be heavily influenced by the back and abdomen.

The markers are specifically located for a reason and it's a good idea to know the reason. However, every system has its own way of placing markers, so we won't cover a specific marker scheme here. Reference your motion capture manufacturer's guides for their suggestion on where to place markers. They must have tested and retested different marker setups until they came up with an optimum setup. We suggest using their scheme to begin with and then changing it to work best for your situation.

Anatomy is even more important when capturing very intricate data such as hand or facial data because a good understanding of anatomy will help you determine how many markers to use and where to put them. The face is extremely complicated, but the hands are as well, and getting good data from either requires both a lot of informed decision making and some trial and error. Never be afraid to try many different marker setups.

When we don't know what we are really getting out of a capture (which is often true when experimenting), we usually put too many markers on. If there are too many markers you will see which ones are redundant and where one marker can be used, instead of two or three markers. The face is an area where we tend to place too many markers. Depending on the complexity of your facial rig and the requirements of your software application for it, you may need more or less markers. However, until you put some markers on a face, track them, label them, and then bring marker data into your 3D animation system; it's not possible to know if your marker setup will give you what you're looking for. It may take three tries to get experimentation to work reliably.

The first try is done with some guesswork on marker placement and we know in advance that we are not going to get perfect data, but it will give us a lot of useful information. We take all the information from the first try and apply it to the second try. This usually gets us 75% there. The third time we narrow in on everything that we want: a marker setup that gives us clean useful data, how markers relate to the facial rig, and how we can get repeatable results (read Chapter 10 for more about facial capture).

So what do you do if you're not capturing a person but an animal? We usually try to take a human marker set and distort it to match whatever we're capturing. A dog is a good example. Keep the back legs as in a human marker set, although you need to change them for the natural bend in a dog's legs. Make the spine come forward out of the hips instead of going up. Rotate the upper arms down to the floor and shorten them. All the while make appropriate adjustments for the differences in dog and human anatomy. This method has a lot of flaws, but gives you a starting point. The biggest flaws are that a dog's back curves differently than a human's and that the shoulders are much more important in locomotion. Besides these are the obvious changes in bone lengths and positions.

Comparative anatomy is the study of the body structures of different species of animals. It helps us understand adaptive changes that body structures of vertebrates (that includes us, humans) have undergone in the process of evolving from common ancestors. At a glance, a horse's knee and a human's elbow seem to have evolved from the same joint of the common ancestor but a knee of a horse's front leg is comparable to a human wrist and a human elbow is comparable to a horse's joint that's right below the rib cage.

Form follows function. What animals eat determines their body structures. Horses, cows, deer, and other herbivores have thick stiff torsos that support their long intestines. They have long intestines because they need to digest plants that are full of fibers. Meats are much easier and faster to digest than plants. Lions, tigers, and other carnivores don't need long intestines and have flexible torsos that allow them to run fast to catch their prey. If you are capturing motions of a cat, you will need considerably more markers on its torso than when capturing a horse. You will be able to reduce your guesswork if you study the anatomy and movements of the animal that you will be capturing.

There are a number of excellent books on comparative anatomy and biomechanics that can give you an idea how different things behave in motion. Just to name a few, *Life's Devices* by Steven Vogel, published by Princeton University Press, 1988, *Exploring Biomechanics: Animals in Motion* by R. McNeill Alexander, published by Scientific American Library, 1992, and *Vertebrates: Comparative Anatomy, Function, Evolution* by Kenneth V. Kardong, Ph.D., published by McGraw Hill, 2002. Do some research.

2.3.3 Capture volume

The capture volume is the amount of 3D space that your motion capture system can "see." When using an optical system, the capture volume never has a nice rectangular or square shape. It's almost always more of a tent shape with higher points being in the middle. If you're using a lot of cameras which are not pointing at the center of the space, then your capture volume can have a shape that is even more irregular. You'll have to check the capture volume before every shoot. There are certainly a number of things to take into consideration when you are setting up your capture space for the first time.

If you're using an optical system, the number of cameras in your system is a major factor that determines your capture volume's size. You want a good mix of camera coverage over a floor. So, if you only have six or eight cameras (Figure 2.1), you'll probably want your space to be as close to 10 feet by 10 feet as possible, although it could be closer to 8 feet by 8 feet.

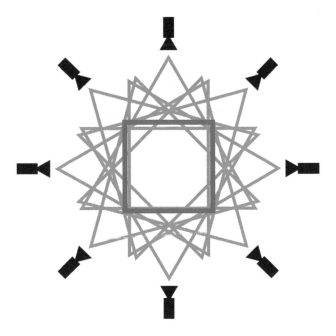

Figure 2.1 *Simple eight camera setup*

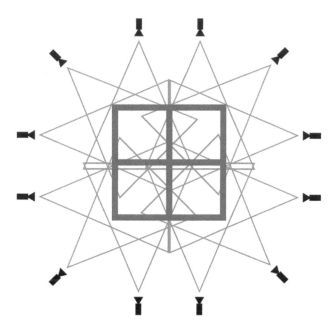

Figure 2.2 *Cameras set up in a zone approach*

If you have more cameras, you may want to create a type of overlapping "zone" approach in which the cameras are grouped and focused on quadrants, instead of all focusing on the center of your mocap space (Figure 2.2), but you should still try to get the center of the space in view of the camera if at all possible. This zoning approach may or may not allow every camera to see the actual center of your capture space depending on the size of your capture volume and the number of cameras that you have. In certain types of capture spaces it is possible that only three or four of your cameras can see the center of the space and other cameras have overlapping coverage areas with them. This type of setup is fine, but you need to check your manufacturer's guidelines on setting up a volume.

Find out the focal length of your camera lenses. The focal length is the distance between the lens and where light converges when the lens is focused on infinity. The focal length controls the magnification and the angle of view; 35 millimeter is a fairly common focal length. If you want to be further away, you need a higher number such as 100, and if you want to be closer, you use a smaller number such as 12. In the case of the higher number, you'll get less light into your camera, and the smaller numbers will create greater lens distortion. Make sure that you choose the right lens for your capture.

Also you want to know the approximate distance between the center of your capture space (or wherever you're focusing your camera) and where the camera should be mounted on a wall or a tripod. However, be aware that a camera lens has a depth of field (an area of acceptable sharpness). A camera lens can focus on whatever is in the depth of field. Hence, if the center of your capture space is within a camera's depth of field, the camera is focused on it. The distance between the camera and the center of your capture space does not have to be the exact distance that you figure out from the focal length.

The first step in roughing in a capture space is to place markers on the floor to outline the area that you're interested in. Next, decide what the desirable height for the edges of your capture volume is and place markers at the height on tripods (poles, c-stands, or anything else that is tall enough). Then, place the tripods on the outline of the capture space where markers are placed in the previous step.

The height of a capture volume depends on what will be captured. If your capture subject will be jumping up on a trampoline, you want to increase the height of your capture volume by reducing the capture space in terms of the square foot of floor area. The height for the edges of a capture space is normally somewhere between 6 and 7 feet.

Finally, adjust each camera's aim so that the markers on the floor and the tripods are seen by as many cameras as possible. Markers in the capture space's corners should be seen by at least two cameras, although three cameras would be more ideal. Adjusting camera aims is easier if you can do it with another person. While you are on a ladder to change a camera len's aim, the other person can look at what the camera is looking at and tell you in which direction the camera aim should be moved. If you have to aim cameras by yourself, you may want to project what the camera is looking at onto a wall so that you can see it from the ladder you are on.

After setting up cameras, you need to test if your capture volume has any blind spots. To do this, walk around outside the capture space with an object with at least three markers on it. With three markers you can track the object and see where the markers go in and out of the capture volume. Make sure to move it all over the space. Thus, you want to move it close to the floor and in the air to find out where markers disappear. Once you know where the blind spots are, adjust the cameras covering that part of the floor to eliminate the blind spot. You may need to calibrate your system first in order to have it compute the trajectories instead of just eyeballing when markers are in or out.

There are times that covering one blind spot opens up others. When this happens you either need to reposition your cameras and try to cover an overall smaller space, or you have to decide which blind spot you can live with. Either way, try to make the most out of what you have, but still remember that the better data you get, the better results you'll have.

With a magnetic system, the size of a capture space is usually defined by the system. Hence, you can't change it depending on specific needs for a shoot. A magnetic system needs to be calibrated for items in the room that may disturb the magnetic field. Calibration must be done in order to create a corrected magnetic field and let you know where the capture space's boundary is. Outlining the capture space on the floor with tape helps you know when you step outside of it.

2.3.4 Shot list
The shot list is a list of all the motions that will be captured in the order that they will be captured with critical information on each shot, such as what talent and props will be needed, if any special

preparations will be required, and how long each shot will take. A shot list should be the culmination of your organization regarding the shots that you are planning to capture motions for. It is generated by breaking down a storyboard as we discussed in Section 2.2.3 (see an example shot list in Appendix A and on the CD).

Your shot list is for you as well as for your talent, client, and project staff. Your shot list should help you understand each shot's requirement, organize people and props, and find the best order for capturing all shots that you need. Also the list should inform your talent of what's expected of them and your client of what you'll be doing.

No matter how great the talent is they will get tired especially if there are a lot of exhausting shots in a row. If your actor gets tired during a shoot, it will show up in captured motion data. You don't want your 3D hero to look tired at some points and perfectly fine at others if the character is supposed to have the same silhouette or pose through all of the motions. If you can, spread strenuous shots throughout the day, but still try to bunch them together in small groups. For example, if you have several run motions, try to keep them together. If you also have several falling motions that are tiring for your talent as well, move the falling motions away from the run motions.

Use the shot list to help you know on what day which talent needs to be at the capture site. You do not want any talent to be sitting around all day long by needing them for the first shot in the morning and the last shot in the afternoon. It is your job to group the shots that require the same talent together as much as possible so that you can economize the amount of time that each talent is required to be there. That minimizes the amounts of your time and cost as well. Major set changes or stunt work also need to be figured into a shot list. For a shoot of 1 week or longer, you'll normally have to write several drafts of a shot list until you get all the pieces fit together. Double and triple check your shot list before you distribute it.

Clients often want to sneak a few extra motions in on the capture day. Work it into your shot list in a logical manner and try to stay on track. If what they want will take too much time, let them know that it will impact the day's shoot, but don't refuse to do it. If they want to add another day's worth of shooting, it would be best to let your producer or whoever handles your negotiations with your client talk with them about the time and financial ramifications of adding to the shot list.

2.3.5 Capture schedule

The capture schedule outlines logistics for a motion capture shoot, such as when and where captures will happen, which talent will be involved, if audio is used, and if special props are needed. For the most part, the capture schedule is an expanded shot list. It lays out the schedule for each day of shooting and informs which talent is needed for which shots.

The capture schedule should clearly indicate when the shoot officially starts, when breaks start and end, and when the shoot ends. Another important role of this document is to let talent know that they need to be at the capture site early enough to get in the mocap suits and have any type of calibration or character matching finished before the shoot starts. In that way preparation time will not reduce capture time.

Be sure to have a lunch break. You can use the time to have your lunch, recalibrate the system, and make any adjustments that the actor's mocap suit(s) needs. Mocap subjects cannot easily slip in/out of mocap suits, but they will eventually have to go to the bathroom. Lunch breaks are very convenient times for that. When they take mocap suits off to go to the bathroom and put them back on, all the markers will shift and it will require you to recapture a T-pose, range of motion, scaling position, or whatever your system needs for tracking.

Some people become uncomfortably hot in mocap suits and may want to cool off by unzipping the suits and getting partially out of them. Try to make your talent as comfortable as possible. If you can't have the mocap space air-conditioned enough, let your talent unzip the suit even if that means you will need to capture another T-pose. Have a positive attitude and consider recapturing a T-pose as an opportunity to improve the quality of your data, instead of an unscheduled task brought upon you.

Always make enough room for any special setups in your capture schedule. If you need to bring in a trampoline or set up a flying rig for stunt work, you have to build setup time into your capture schedule so that everyone will know what needs to happen next and also setup time won't cut into shoot time.

You want to follow your capture schedule as much as possible, but remember your capture schedule is a guide. Capture schedules are not set in stone and everyone appreciates a flexible approach, especially if they are new to motion capture. So, if you have to deviate from your schedule, try not be stressed out about it. Be flexible.

2.3.6 Rehearsals

Rehearsals are vitally important to the smooth running of a motion capture shoot. By rehearsals, we mean that the actor(s) will rehearse at an off-site location but not at the motion capture studio. There are many reasons to have off-site rehearsals, so let us go into just a few of them here.

Time is money in a mocap studio. Each time you start using a mocap studio, someone has to pay for the electricity, the salaries of the staff who works there, and the wear and tear on the computers, cameras, mocap suits, and markers to name just a few of the obvious costs. Your client does not want to waste any part of a day in a mocap studio. While a director or choreographer is figuring out how to approach a shot with talent, your client will be paying for the mocap staff who is not capturing anything. That will be a large waste of money on the client's part. Rehearsing at an off-site location can certainly avoid such a situation.

Some clients may swear that they are prepared although they're not. Talk to your client about having a rehearsal at an off-site location. If you can convince your client to have a rehearsal, it will benefit you as well. You will be able to flag difficult to capture motions and foreseeable problems.

Your clients may never give you a shot list or will give you one that will change a lot when they start shooting. Rehearsing will help the clients to better understand what they are really looking for. It also lets the clients know if they want something that they have never thought of before until they see it acted out in front of them. Rehearsing will help your clients to create a better shot list.

You can't force your clients to have a rehearsal. If you tell your clients that you want them to have a rehearsal because it will make your life easier it will be very unlikely that they take you seriously. They may tell you that they are not paying you to have an easy life. On the other hand, if you explain to your clients that a rehearsal can save their money and time, they will be more agreeable. If you are the client, then you can of course make the people whom you work with have a rehearsal and get ideas about how the mocap shoot will go, how much you can possibly capture in 1 day, etc. Planning ahead and thinking things through always make the whole mocap process easier.

Many times clients will not know exactly what they want to do because they have not done motion capture before or a member of their production crew is not familiar with it. There is also a case where the clients have done it several times before and know what to expect, but they are not used to your way of doing things. You will encounter individuals who have seen mocap on YouTube and assume that they know how it's done. It is your job to educate them about how motion capture works and how your studio should run in a non-hostile, cooperative manner. Also be flexible enough to make them comfortable by relaxing your methodology and adding a little of theirs. You'll find out you get much better results when everyone understands what other people are doing and why. You'll also make your methodology better.

2.3.7 Props

One area that people tend to work on at the last minute is the props. There are a variety of good and bad reasons that props are added at the last second, but you normally have to accept it and try to incorporate them into what you're doing. By props we mean "markered" props that will be captured. They should be regarded as actors. The reason to think of them this way is that you will be capturing their motions and they will be moving in relationship to your actor(s).

Because props are usually moved in relationship to the actor(s), special consideration needs to be given to where to place markers and how many to place. You usually need at least three markers, but think about using more to help with occlusion problems. Occlusion will normally happen if the prop that you are using is picked up by an actor or handed from one actor to another. Unless you know for sure that the markers will always be seen by the cameras, it's best to double up on the markers. More about props and markers will be covered in Chapter 6.

When you are deciding where to place markers, think about where the prop will be handled and what other markers on the actor will be in the same vicinity. If the prop is a rifle, the butt of the rifle will be up against the actor's shoulder. In this case, you wouldn't want to put a marker on the butt of the rifle (Figure 2.3), but a few inches up on the top side of the stock would work well (Figure 2.4).

Figure 2.3 *Improper marker placement*

Figure 2.4 *Better marker placement*

Another consideration for props is their relationship to other props that are captured on the set. If you are capturing a sword, you will of course need to pay attention to the relationship between the sword and the actor's hand that's holding it. If the sword hits another sword, you will also need to pay attention to the relationship between the two swords. You do not want markers to get flung off when the swords collide.

Think about these relational issues when placing markers on your props. Test capture your markered props to see if marker placements are working and also if props are reflective enough to cause an issue with your motion capture sensors. If a prop is too shiny, then ask the prop's owner if it is OK to cover the prop up with gaffer's tape or paper tape. Also, before you start attaching markers on a prop, talk to the prop's owner about potential damages to the prop. Some tape or glue will take the finish off of a prop's surface and damage the prop.

When you attach motions to a 3D character, a prop should not be treated as a natural rigid extension of the character's hand because it is not. When people use hammers, golf clubs, etc. the items will swing and move in their hands. Thus, a grip, or how an actor holds a prop in the hand, can change in the course of an action. That's why props cannot be regarded as rigid extensions of hands. There are a few different types of grips, but let's look at two of the most common ones. One is the grip that you use when you are pointing a prop at someone. It is more of a golfing grip where the thumb is open and the object lies diagonally across the palm. The other is the grip that you use when you are holding a prop upright. It is more of a hammer grip where the thumb is normally closed and the object lies along the palm parallel to the knuckles. There are obviously other ways to grip items, but these two can exemplify the point that an object moves within the hand and changes its orientation with respect to the hand. Do not simply parent a prop object to a hand of a 3D character. Spend time making sure that you properly track the props and the actors and then learn good editing techniques to take the above fact into account.

A very useful technique for editing the relationship between a prop and a hand is the use of what MotionBuilder calls "Auxiliary Effectors" or what Nuance calls "markers." An auxiliary effector (marker) can be attached to one object and then influence another object. You can create one that is attached to a prop and have it influence a hand so that the hand moves a little to compensate to the prop's movement but not a lot. Locating an auxiliary effector (marker) close to the web of the hand between the thumb and forefinger usually works well (Figure 2.5). This method does not work for every conceivable motion but provides a lot of flexibility in editing the relationship between a prop and a hand. It is a good place to start.

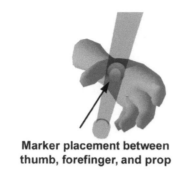

**Marker placement between
thumb, forefinger, and prop**

Figure 2.5 *Object–hand relationship*

2.3.8 Suits and markers

In the preproduction stage it is important to prepare markers and mocap suits to which markers (or sensors) will be attached. Be sure to have more than enough markers of the right size and enough mocap suits and shoes that fit the capture subjects. Markers, suits, and shoes should be clean and in good condition. Repair or replace damaged ones. On each capture day, before the talent and clients arrive, have the suits, shoes, and markers set out and ready to be put on.

Let's look at each item with a little bit more detail.

You never want to run out of markers. To be sure that there will be enough markers, make an estimate of the number of markers that you will need and add about 30% to the number. Have that many markers in a usable condition. The extra markers can be a lifesaver in case some markers get lost or damaged or there is a last minute change. Marker surfaces can get oils from our fingers by excessive handling and also get scratched or rubbed off by anything that comes in contact with them. Both cause marker surfaces to be less reflective. Add more reflective tape if this is the case.

If small optical markers will be used for facial or hand tracking, prepare a visual guide to show where the markers need to be placed. Pasting tiny markers to the capture subject's face with glue and tweezers is time consuming and the placement is critical to getting good, repeatable results. If you are going to do several days worth of facial capture, create a facial mask with small holes that indicate the locations of markers. Place the mask on the actor's face, mark the positions of the holes with an eyeliner pencil, remove the mask from the face, and paste markers on the marked spots. The mask allows you to use the same marker placement every day. A mold taken from the actor's face gives the best result.

Make sure that the suits are properly laundered, their zippers work, and there is no tear in them. You want your talent to wear a suit that is tight rather than loose. That can stress the suits, especial seams, and some get torn, although mocap suits are made of stretchable materials.

Velcro is normally used to attach markers to the mocap suits. Stock various types and sizes of Velcro pieces to attach markers to the suits and repair any problems with the suits.

Shoes, headbands, and wristbands can be considered as part of the mocap suit since markers will be attached to them. You may choose to attach markers to them permanently and use them as "pre-markered" shoes, headbands, and wristbands.

Try to get measurements of your talent beforehand. If you are using professionals, they will be able to give you all the relevant measurements. Make certain that there will be suits, shoes, headbands, and wristbands of appropriate sizes waiting for them on the capture day.

That's about all we can think of for preproduction. Everyone approaches preproduction in a slightly different way but it is always important to figure out every detail as much as possible in the preproduction stage. We are now heading to the next chapter on creating a pipeline which is closely related to what we have discussed in this chapter.

3 Pipeline

Various types of information that contribute to a project move through a production pipeline. The information changes its form while moving from one step to the next (e.g., a script turns into a storyboard; marker data into skeletal data). In a large production environment, teams of specialists are involved in the production. Each team is responsible for a section of the pipeline. In a small production environment, a single individual or a small number of crew members are in charge of the entire pipeline. No matter what size a production environment is, it is crucial to have a well-designed and well-tested pipeline that projects move through smoothly.

A project pipeline starts with preproduction and ends with post-production. Appendix B shows the middle part of the pipeline that is being used at the University of Texas at Dallas and the Ohio State University as of 2007. The diagram starts with calibrations (in the upper left corner) and ends with rendering. It is an example of a mocap production pipeline that consists of a number of steps, software applications, and file formats. Depending on the type of projects that you will work on and the hardware and software that you have, your pipeline may look different from the sample.

In this chapter, we will discuss key points for designing a pipeline. We will skip over preproduction here since it has been covered in Chapter 2 but the preproduction phase is the first and the most important step in any pipeline.

3.1 Setting up a Skeleton for a 3D Character

The character in a game or animation is ultimately where you want the mocap data to go to. The skeleton that goes inside the character is what you will work with in your motion capture system. There are quite a few different ways to look at the skeleton.

The first questions are "What is going to be done with the skeleton and at what level?"; "Is the character going to be primarily key-framed with some motion capture?"; "Is the character going to be all motion capture?"; "Will any editing be done in a motion capture specific package, such as MotionBuilder or Nuance?"; and "What does my final output format need to be for my client?"

There are times when the skeleton needs to be built in a certain way specified by the mocap software. Check to see if the software you are using expects a specific stance or the joints oriented in a specific way. One of the old conventions is letting the y-axis of a joint point to the next joint (i.e., face down the length of the bone between the joint and the next joint), the z-axis point forward (i.e., toward you if you look at the skeleton from its front), and the x-axis be the cross-product of the y- and z-axes.

Some riggers use a similar approach but may let the x-axis of a joint point to the next joint, instead of the y-axis. Consider these differences before you get too far into your process. One area where you may ignore the rule about the joint rotation axes is the foot or ankle joint. We usually make this one orthogonal to all the other bones so that the joint's z-axis still points forward. This can make editing and rotation of the foot easier.

Make sure all the joints have a major axis consistently facing down the length of a bone. These days the x-axis is often used as the major axis but the y-axis was most often used in the past. The spine is one of those areas where you may map one variable to several joints. If you do not orient the rotation axes of the spine joints consistently, the joints will bend in inconsistent ways.

There are multiple ways to handle the spine. How many bones should be used? A good rule of thumb is that four will work OK but seven or more is better. One is never acceptable and two will just barely work. The back can curve in more than one direction along its length. Try this: bend over to touch your toes. Isolate your back and try to bend your chest and head up to look in front of you. You will notice your lower back is in almost the same shape but the upper back is now in a much different shape. Trying to replicate or track every vertebra is usually not feasible. So try to give the skeleton enough joints to allow the mocap software enough flexibility to get the best solutions it can. A bit of trial and error is usually involved.

The shoulder joints are unique because they can translate as well as rotate. They have a wide range of motion, which makes them more prone to dislocation than other joints. Most animation software is not able to handle the shoulder joints properly and treat them as three degrees of freedom (DOF) rotational joints. However, there are locations that the arms cannot reach without the shoulders' ability to translate and rotate. An additional joint can be created between the shoulder joint and the spine to track the rotation and translation of the shoulders.

Generally the character is already designed and modeled before the shoot. So, the proportions of the 3D character have already been determined before the data is captured. The best results come from a skeleton that matches the proportions of the person in the suit as closely as possible. If you can use the proportion of the capture subject, take the marker data in a T-pose (Figure 3.1) and use that as a basis for the lengths of the skeletal segments (see Section 4.2.2 for more details).

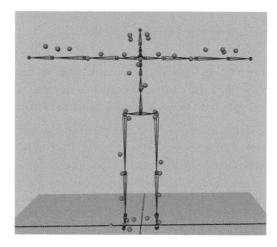

Figure 3.1 *T-pose*

Think about where the editing will happen. Let's suppose that your skeleton has finger joints for key-framing but you are not going to capture any finger motion. Your mocap skeleton does not need the finger joints unless you plan to add the finger motion inside the motion editing package. If the editing (key-framing in this case) will happen in a 3D animation package, then leave the finger joints out of your mocap skeleton since they will never be touched and there is always the possibility of accidentally mapping data to them.

3.2 Calibrations

The first type of calibration is preparing the mocap system so that it can tell where all the sensors are. To do this with an optical system, you usually have to first tell the cameras where they are and correct for as much lens distortion as possible. With a magnetic system, you would be making corrections for other electromagnetic fields that may be interfering. The second type of calibration is for the person(s) and/or object(s) being captured.

3.2.1 System calibration

System calibration is usually pretty straightforward. It normally involves some type of calibration device, such as a wand that has some pre-defined markers on it. These markers are placed at a set of distances from each other so that when the software sees the markers in one camera, it will be able to tell what is which. When multiple cameras all see the markers, the cameras will start to figure out where they are in relationship to the other cameras (Figure 3.2). Once the cameras know where they are, they can triangulate a marker that two or more cameras see.

Some systems also have a secondary device that sets up the world coordinate system's axes and origin. The world (global) coordinate system is initially set by one of the cameras and all the other cameras' locations and orientations are computed from that. The coordinate system setup allows you to decide where the world coordinate system's origin should be, what type of system (e.g., y-axis up) it should be, and which direction each axis should point at.

Remember that once your cameras are set up, you do not want them to move at all. Something as small as a constant vibration in a wall can move a camera. The camera would be skewed from its real position. When it is used with the other cameras to determine the positions of markers, the system will generate bad data. Thus, the slightest movement of a camera can affect data greatly.

Mount your cameras up on solid walls or grids if possible. If they are on tripods, create barriers around the tripods to keep people from accidentally leaning on them or bumping them. Place no camera close to a client area or a high traffic area such as between your control room and your mocap space. Always remember that even something like a slamming door can shake a camera and cause the camera to fall out of calibration.

The better your calibration is, the cleaner the data will be when it comes out of your system. Calibrate in the morning, lunch, and at the end of the day. If you are using an optical system make

Figure 3.2 *MoCap calibration wand*

sure in the morning to allow the cameras enough time to warm up (usually 10–20 minutes). Having three different calibrations per day can help, especially if a seemingly insignificant amount of displacement creeps into your mounted cameras' locations. Even the afternoon sun can move your cameras by heating up the roof of your building, making the roof expand, and warping the building structure to which the cameras are attached. It is also a good practice to check your data whenever possible just to see what kind of quality you are getting. Remember that once the calibration starts to go bad, there is not much you can do with the data unless you stop everything and recalibrate.

3.2.2 Subject calibration

Your capture subject is whatever you are interested in capturing. It can be as mundane as a hat, or as complex as half a dozen people performing extremely complex interacting motions.

If you view most mocap systems as initially totally ignorant of what they are tracking, that is the right approach. You need a way to tell the mocap software what it is looking at and what the relationships among the markers are. This is done with a subject calibration.

The subject calibration normally takes place with the subject being in a T-pose, the subject exhibiting the range of motion (i.e., maximum flexibility) of each joint, and a generic skeleton (or a template). Let's look at each.

In a T-pose, a subject should stand facing down an axis of the mocap space. We commonly use the positive z-axis but different systems have different axis orientations. The feet are about shoulders' width apart and face as directly forward as possible. Arms are straight out to the sides with palms down. The head should be facing forward and level, not looking up, down, left, or right, but straight forward. The body should resemble a T (Figure 3.3).

Figure 3.3 *Subject with markers in a T-pose*

A shot of the capture subject going through each joint's range of motion is called a "range of motion trial." The software uses a T-pose and a range of motion trial to correctly identify and track the markers on the subject, and create better data. The range of motion that you ask the subject to do is as simple as moving each joint through all the DOF. Have the subject do that for all the joints that will be tracked, including the spine and the neck. If you are not going to track fingers, there is no need for the subject to do range of motion for the fingers. It is a common practice to have the subject start and end the range of motion trial in a T-pose.

Once you have the data of a T-pose in the optical system, identify the markers. A label (or an individual name) should be assigned to each marker. The labels are used to create a statistical model that your software will later use to identify and track markers on its own. Also the labeling step gives you a chance to check that all the necessary markers are in place and in their proper locations. Label all the markers including the ones that drop in and out due to occlusion. Discontinuity is common in the waist area when the subject bends a lot. The markers on the front waist are often occluded by the thighs bending up against the stomach.

When all the markers are labeled, run the subject calibration process. The process fits your system's generic skeleton to the labeled marker data of your capture subject in a T-pose in a range of motion trial. Different systems have different methods for this but the main concept is the same. First, the following is given to the process as input: a generic definition of a skeleton, the marker and skeleton relationships (i.e., relationship between each marker and the skeleton segment that the marker is associated with), a T-pose, and a range of motion trial. Next, using statistical measurements and comparisons on given data, the marker and skeleton relationships in the generic skeleton are optimized to best match the actual locations of the markers on the subject. The proportions of the generic skeleton are changed to match the physical dimensions of the captured subject. Thus, the subject calibration process stretches or shrinks and rotates the generic skeleton's segments so that the skeleton fits the capture subject's proportion and posture. The T-pose is important in this process. So, have the best T-pose possible.

If you are going to capture objects, it is very important to record their placements. Either take reference photos or keep notes. Develop placement rules that you always use when you place objects. Rule 1 may be making the marker closest to the capture subject the beginning of a prop and the marker farthest away the end of the prop. Rule 2 may be placing the beginning of the prop at the center of the capture space while the end of the prop points at the positive z-axis. If you use only three markers on a prop (although we say four markers are minimum and six markers are recommended) it will be hard to tell a bat from a sword or a walking stick and which end is which when you look at data. If you do not keep track of these seemingly minor things in a prop heavy environment, they will drive you crazy. Always keep good notes and think about how you will know the orientation of the object in every shot. It is a good idea to use a vinyl tape to outline an object on the floor and replace the object inside the outline at the beginning of every shoot.

If you are going to capture a prop that has no joints and no, or little, flexibility, such as a bat, you do not need to capture any range of motion or track it as a multi-jointed object. If your prop has a lot of flexibility, like a golf club, you can create a skeleton for it and track it as a multi-jointed object. If you have a prop with jointed sections, such as nunchakus, create a skeleton, capture its range of motion, which will create a better statistical model, and track it as a multi-jointed object (read Chapter 6 for marker setup for props).

3.3 Capture Sessions

Capture sessions are very reminiscent of being on a stage or shooting live action. Everyone has to know when to start their part of the process and when to stop. A capture schedule should be created and followed. It should have a complete set of times for T-poses, changing props, changing camera setups, camera calibrations, lunch break, and any other necessary business. It also should dictate who and what will need to be on or off the mocap stage.

3.3.1 Audio and video references

It is usually a very good idea to make both video and audio reference. A video camera can be set up in the corner of the room. This camera can also be used as a "slate" camera. Using a clapboard (or something else large enough to be written on) the slate camera lets you record what shot you are on and which take of the shot you are currently capturing. The camera should be in a fairly generic location where you get good coverage of the mocap area but not in the way of the mocap crew or the subjects being captured.

For video and audio, always get the best resolution that you can. If you are recording to tape, be sure to have extra tapes with you that are already dated and numbered. Also always have a few blank tapes lying around just in case. Before the shoot starts check your video equipment and make sure it records both video and audio. Use a tripod. Do not shoot by holding the camera. Avoid excessive zoom in and out.

It is possible that audio will be played during a shoot, such as music for dancing or a sound track that people will act to. Take an output of that and bring it into one of your audio channels so that you have a clean recording of the audio. Use the other channel for live recording so that later you can hear onset comments and concerns. Just as a note, always pretend that you are being recorded and assume that your boss and the client (or your professor) will listen to everything on the tapes. It is possible that once you get comfortable with the mocap process and the people around you, you may say something you do not want to be on tape, so, watch out.

A note about time code: If you can successfully push time code onto your video track, audio track, and into your mocap data from a common source, do it! There is no reason not to use time code but you will need a dependable time code generator and a reliable way to feed it into your video/audio system and mocap system. If you are going to have synchronized time code in audio/video and mocap data, make sure it will not crash the mocap system or "hiccup," causing time code to be no longer in synch across applications.

3.3.2 Organization

Having everything organized and everyone informed about how to be ready for each shot helps the shoot go well. The capture schedule (detailed in Section 2.3.5) for the day should be one of your guides, and everyone in the production should have a copy. There are usually last minute changes, so be prepared to run new copies every morning and hand them out. Any changes should be highlighted or explained so that the production staff and talent know what to expect.

Make certain that your props are in order and in a logical place. If you are going to be changing props out a lot, do not leave them in another building or on a different floor. Keep them on set if you can. Cover them up so that the mocap cameras will not see markers on them. Having markers and mocap suits ready for the day's capture is also important.

Get into the shoot early on the first day and have everything turned on. It is good to let everything warm up and calibrate, even if no one has shown up. It is one less thing to do. If you find a damaged or broken cable, or some other problem, you will have time to fix it before everyone else shows up.

Have an "emergency" cart in the mocap space. It is a storage cart that contains supplies for repairing or replacing markers and various types of tapes, such as reflective tapes, gaffers, and double-sided toupee tapes. (We recommend double-sided toupee tapes for contact with human flesh. It is durable and sticky.) Check if everything (e.g., tapes and glues) that comes in contact with human skin is hypo-allergenic. Even if the packaging says so, it never hurts to test it on someone just to make sure. Some people have horrible reactions to latex and some of the tapes have latex in them, so pay close attention to them.

If you are having talent brought in or are having a large group of people, make sure that they are taken care of. Provide a light breakfast, lunch, dinner, or whatever is appropriate. Always have plenty of water on set for the talent as well. Take a little initiative and see what type of snacks or candy the talent may like in order to give them an energy boost or just to allow them to munch on something they like in between takes.

3.3.3 Preventing occlusions

With an optical system some occlusion is not preventable but making the effort to reduce occlusion will result in cleaner data. Let's look at some ways to prevent markers from being occluded.

One common way occlusion happens is having too many capture subjects in a shot. Many times there are too many capture subjects in a shot who are not interacting with each other. If it is possible to break the shot up into several shots with fewer capture subjects, do it. If they all need to be in a shot together for interaction or timing, you will just have to spend extra time in post-processing.

Whenever possible, ask capture subjects not to put their hands directly over a marker. This is not always possible but asking them can help in the end. Another thing to avoid is impeding the performance of the person being captured. When you work with dancers, do not give them too many rules. Just let them perform. Having to think about their performance and all the rules you gave can be distracting. The same goes for anyone who is trying to create a good performance. They may position their hands in certain ways that make the mocap crews cringe because of all the occlusion. However, forcing performers to restrict their performance will cause the performance to suffer. It is balancing good performances with avoiding occlusion.

If the motion you are capturing is fast paced with lots of action, or if it is for background characters, you probably can get by with some occlusion since it is not the focus of what you are creating. If a particular action will be the focus of the final camera shot, work to reduce any possible marker occlusion. If occlusion persists, try variations of the motion.

Props can be made more marker friendly if they are created as frames and not as solid pieces. One example is a table. If a capture subject is going to only lean on the table, you do not need a table-top. You only need the frame of the outside edge of the table. If a capture subject is going to place the hands down on top of the table, there are a couple of options. One option is to use a very loose wire mesh over the top of the table. The other is to use a platform (such as a tripod) that has a board attached to the top at the same height as the table and in the position where the subject will make contact. The reason to do this is to have as little physical material as possible between any markers and the cameras. Normally cameras are placed above the mocap space looking down. If a capture subject is next to a regular table with a solid tabletop, the table will block some of the cameras from seeing the subject's legs and feet. Minimizing this blocking helps prevent occlusion. Every little bit of extra clearance helps and having just one more camera see a marker can be the difference between usable and non-usable data.

3.4 Cleaning Data

The majority of systems write out marker data and some systems write out skeletal data. Both need to be cleaned. Data cleaning both is basically the same. Either there is bad data that you need to remove and replace with better data or a best guess, or there is no data and you will need to create something.

As mocap systems get better, these kinds of problems are disappearing. The software is getting better at guessing what an occluded marker would be doing based on other markers around it and filtering out low- and high-frequency noise. But until these techniques become 100% reliable, we still have to clean data and spend some time trying to fix what should not be in the motion.

The most common problem in the data cleaning is gaps in data. You may let the software figure out the best way to fill in gaps in the data but inspect what it does. The software results are usually acceptable but when they go wrong, they can go very wrong. So, when the result is not acceptable, try another method, then another. As a last resort, try fixing the problem by hand. Place the data where it is supposed to be. Exhaust all the other options before doing this because it can be quite time consuming. Check your software and see what kinds of tools are available.

Another problem that often shows up in mocap data is spikes. If you look at skeletal or marker data, you may find values that form sudden peaks or suddenly shoot off in a strange direction. The best solution is to simply chop those sections out or use your software's filtering algorithms to take the spike out. With Euler angles (which will be discussed in Chapter 12) it is sometimes difficult to take out spikes, but again, the software should be able to handle quite a bit of that for you.

One common problem in mocap data is "shaking" when you need someone to be standing still. It seems that a group of markers are basically staying in the same area but slightly shifting around in every frame. There are numerous reasons for this and there are also several approaches to solving it. One option is purposely creating a data gap by deleting the shaking data and filling in the gap with an interpolation method. For instance, if you know that a foot is on the ground and should not be moving for a certain period of time, delete all the shaking data of the foot in the section and perform an interpolation between the values at the beginning and end of the data gap.

Another option is reconstructing the trajectories of the affected markers in the section by changing parameters for the reconstruction algorithm. The markers may become stable on a second pass. Yet another option is to run a filter across the affected markers.

The level of detail you need to clean your data largely depends on the type of product your final product will be. If you are working in film or broadcast, each shot will be rendered using a camera pre-determined in the preproduction phase. You cannot change the camera freely. If you did, the continuity would be missing from the final product. You need to edit the characters' motions based on how they look through that camera. That is often an advantage. For example, suppose that the mocap data of a character's hands is bad and the hands is bent sideways. If the camera does see the hands, there is no need to fix the data.

In a video game and other interactive media, the player (or user) has full control of the camera and can look at the characters from any angle. The player can navigate through the entire environment and get close to any characters. You have to clean the data from top to bottom and pay enough attention to the background characters' motions as well as the main character's. There are no places you can cheat. The one luxury with games is that shots are normally very short and have a single character. There are "full motion videos" or cut scenes with multiple characters and complex motions but the player does not have control over the camera in these pre-recorded animations. We will discuss cleaning data in more detail in Chapter 4.

3.5 Editing Data

There are two types of data to edit. One is marker data, and the other is skeletal data. Because marker data is translational data and skeletal data is translational and rotational data, translations and rotations need to be dealt with differently depending on the data type. We will look at both types after a brief talk about hierarchies.

The hierarchy is a system of relationships among elements where each element is a subordinate (a child) of a single dominant element (a parent). The element at the top of a hierarchy is called the root. Each element (except for the root) has one parent and an arbitrary number of child elements. A transformation applied to a parent is applied to its child as well but a transformation applied to a child is not applied to the parent. Geometries, markers, and joints are often organized in hierarchies. For example, suppose that there are two entities, A and B, in 3D space and A is parented to B. Thus A and B are in a parent–child relationship in a hierarchy. If you translate B by 5 units, A will also move with B 5 units. If you move A 5 more units, B will stay in place and will not move. (See hierarchy movie on the CD.)

Marker data is normally organized in a hierarchy in which all of the markers are children of the root and none of the markers are parents of other markers. It is a flat hierarchy (Figure 3.4).

In a skeleton, which is a hierarchy of joints, the root is normally the hips. Building off the hips are the legs and the spine. The spine goes up until the neck and shoulders/arms branch off. It is a more complicated hierarchy than a marker data's hierarchy (Figure 3.5). It is often beneficial to have a node above the hips, called a "reference," as the root, instead of having the hips as the root. This allows you to make changes on the whole skeleton without changing the transformation values of the hips or any other joints. For example, if a skeleton is placed and keyed at a desirable location in an environment using a reference node and if mocap data is imported onto the skeleton, the skeleton will keep its location keyed in the reference node.

Figure 3.4 *Hierarchy*
of a marker data set

Let us start talking about editing marker data. If you are using an optical mocap system, marker data is a collection of the x-, y-, and z-coordinates (or x-, y-, and z-translation values). None of these are dependent on the others, so if the x-coordinate is changed, it will not affect the y- or z-coordinate. For instance, let us suppose that the y-axis is the vertical axis and that we want to move all the

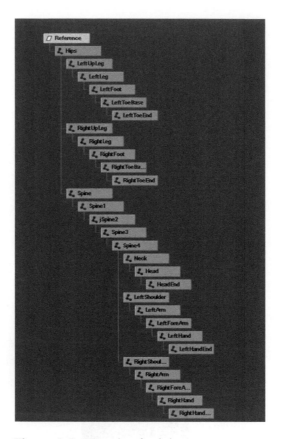

Figure 3.5 *Hierarchy of a skeleton*

markers up 5 feet in the air in order to have a character walk on a 5-foot tall platform. Select all the markers of the character and add 5 feet to the *y*-coordinates. Then the markers will be 5 feet in the air. A much better way to do this is to select the root node that is the parent to all of the markers and add 5 feet to the root's *y*-coordinate. Changing one *y*-coordinate (the root's *y*-coordinate), instead of multiple *y*-coordinates (all the markers' *y*-coordinates) moves all the markers up 5 feet.

An issue that you may face is translating a character in one continuous shot. For instance, a character is supposed to walk and then step up onto a sidewalk but the stepping up action was never captured. The action needs to be created by editing walking data. This is a challenging case.

If a vertical translation is added to the whole character at the moment it is supposed to step up, the character will look as if it were magically lifted up. If blending alone is used, the character will look like it's on an escalator. For a better result, apply both translation and blending. Add emphasis to the hips and chest so that the character will show the effort of pushing off the ground to get up to a higher level. Another approach is to use mocap data to rough the motion in and then key-frame the stepping up action building in the right level of effort.

Now let us use a joint in a skeleton to think about editing rotations. Rotations are more difficult to work with because we use Euler angles. The x-, y-, and z-rotations of a joint are not independent from each other like the x-, y-, and z-translations. Thus, if you change the y-rotation, it may affect the x-rotation or the z-rotation, depending on the order of rotations (see Chapter 12).

If a character's arm needs to rotate to a certain position, the action added by data editing should move the arm to the position in a realistic manner. A linear interpolation from the original rotation angle to the new rotation angle makes the action mechanical and unnatural. Apply slow in and slow out and make the action look more fluid and natural. Also let the action take place in an appropriate amount of time. If the action takes too long, it will look like slow motion. If it is too quick, it will also look wrong. Keep trying different timings until you find the timing that looks best for the specific action.

Since a skeleton is a set of joints in a hierarchy, rotating a joint in a skeleton rotates all the descendent joints (the joints that are below the joint) in the hierarchy, for example rotating a shoulder joint will move the upper arm, lower arm, hand, and fingers. Move one of your arms above your head and see how it affects your balance. Try that in a neutral balanced position (e.g., sitting down) and a precariously balanced position (e.g., standing on one leg). You will notice that if you move a character's arm, you need to adjust the character's balance. If the spine is rotated, it will move the head and arms as well. Counter rotate the head and arms to correct the character's balance. Editing skeletal data will be detailed in Chapter 5.

Motion editing is the closest thing to animating in the mocap production pipeline and a fine eye for motion and balance is needed to make the editing results look believable.

3.6 Applying Motions to a 3D Character

When mocap data is cleaned and edited, it is time to apply the data to a 3D character. This means that you need to define how data relates to the character's skeleton. There are multiple ways to do it. Let us look at data application methods for both marker data and skeletal data.

Let's start with marker data. Marker data requires you to assign markers to the joints in a skeleton directly or indirectly using an intermediate skeleton. An example of the direct approach is bringing marker data into Maya and applying it to a skeleton there. An example of the indirect approach is bringing marker data into MotionBuilder and applying it to a skeleton via an intermediate skeleton, called an Actor. (The former example will be detailed in Chapters 6, 9, 10, and 11. The latter example will be detailed in the second half of Chapter 4.)

If you are applying marker data to a skeleton either directly or indirectly, you need to let your application know which markers are associated with which segments of the skeleton (i.e., which markers move and rotate which joints). First, create a character's skeleton in a T-pose and match it to the mocap data of your capture subject in a T-pose. At this point, the skeleton is probably larger or smaller than the marker data. Resize the skeleton if you are using a direct method without an intermediate skeleton. If you are using an indirect method, scale and adjust the intermediate skeleton to deal with the size difference. Secondly, let the application know which marker should move which segment of the skeleton. And finally, retarget the motion to the skeleton of your final 3D character. If you try to fit marker data to the skeleton and tackle retargeting issues at the same time, it can get too complicated, especially if you are new to mocap. Try to conquer one problem at a time. Take a layered approach.

Let's look at skeletal data. Skeletal data is applied to the skeleton of a 3D character directly. There are several issues to consider in order for this to work. Before we look at some of the issues, let us call the skeleton in the skeletal mocap data the "source skeleton" and the skeleton of the 3D character the "target skeleton."

The first issue is the joint names. Each joint in the source skeleton and the corresponding joint in the target skeleton should be identically named if you simply want the skeletal data to be imported onto the target skeleton.

The second issue is the local rotation axes of joints. If the local rotation axes of a joint in the source skeleton and the corresponding joint in the target skeleton are oriented differently, the joint rotation will be messed up. The local rotation axes of each joint must be oriented identically across the source and target skeletons to make any sense out of the motion.

Sometimes these issues have to be solved in the mocap software and other times it is easier to create simple utilities (e.g., scripts) to take care of them. When these issues are sorted out, the retargeting issue still remains. Retargeting is a very common problem and usually seen as a normal part of the process. It can be frustrating at times, but with practice it becomes easier and easier.

Retargeting is making data of one skeleton work for another skeleton of a different size or proportion. For instance, if the data of a capture subject (source) is applied to a 3D creature (target) with long legs and a short torso, the creature's feet will not fall in the proper place. You need to "retarget" the feet so that the footfalls match. Scale the source skeleton up until the source and target skeletons have the same length between the waist and feet bottom. There are times when you want the hands to match up more than the feet. In that case, hands should be the focus of your retargeting. There is no single strategy that works for every case and different strategies are required for walking, crawling, hands important, feet important, and other cases. Retargeting will be discussed more in Chapter 5.

When setting up a pipeline, check if the data type that comes out of the mocap system and the data type that goes into your 3D package match. Unfortunately, you usually have to make the mocap system happy and use the type of skeletons that it spits out unless you have a more advanced system that will allow you to import or tweak joints from another 3D package.

3.7 Rendering and Post-production

Once all the motion is applied to 3D characters and the editing is finished, it is time to look at the motion by either loading everything into the game engine or rendering all the frames using the camera that has been set up in a 3D application based on cinematographic decisions made in the preproduction stage.

It is nice to see the motions you worked on inside the game. Depending on the software and pipeline, motions are loaded into a game several times so that the designers, directors, and producers have a chance to see how motions look and how they transition. This gives you feedback on what will need to be changed.

When frames are rendered in a 3D application, a virtual camera "films" the shot. The camera should be set up for each shot based on the storyboard, animatic, or progression reel, whichever has the latest shot decisions. Rendering generates a still image for every frame of a shot. Each still will have a name and a number so you can keep track of which shots and which frames have been rendered.

Before you render all the frames of your project using the full resolution (i.e., the resolution of your final product), do a pencil test. A pencil test is rendering frames using a lower resolution (e.g., one half or quarter of the final resolution) to test the camera, motion, texture, lighting, special effects, and other elements of your animation. If you plan to use time-consuming rendering algorithms (e.g., ambient occlusion and subsurface scattering) for your final rendering, use a faster rendering option and a smaller resolution for the pencil test. This may seem like an extra step that you can skip but it often keeps you from wasting a lot of rendering time on full-size images that you will need to re-render. You will almost always find something that needs to be fixed in your first pencil test. Repeat pencil tests until everything is worked out for the final rendering.

In the preproduction stage, think about the aspect ratio. High-definition (HD) TV screens with 16:9 aspect ratio are replacing standard definition TV sets with 4:3 aspect ratio at homes and businesses. If you plan to use 16:9 aspect ratio, create a storyboard using 16:9 aspect ratio. Look at a motion using 16:9 aspect ratio. The 16:9 aspect ratio leaves more space on both sides of the screen that you can fill up with extra characters, props, or a background. Use the selected aspect ratio for pencil tests and the final rendering. Be prepared to capture everything you need for the project before you go into the mocap space.

When you come to the final rendering, render in layers and passes. Layers and passes are compositing issues and do not deal directly with mocap but can play important roles in reducing rendering time and improving final image quality. If you can render each character in a separate layer and each attribute (e.g., diffuse reflection, specular reflection, and shadow) in a separate pass, do so. For example, if your main character needs more backlight to separate him from the background, you may need to re-render (or simply alter) one of the passes of the character, instead of re-rendering everything in the shot. In this way you will have more control over your final image quality and be able to create a more visually compelling piece.

Once the frames of all the shots are rendered, bring the sequences of images into a video editing application and compile them into a single image sequence that has all the scenes and shots. Then synchronize it to the audio tracks. This can be tricky, so try to use an easily identifiable sound or the start of someone talking to help with matching the data. The mouth starting to open with a voice is usually a decent place to start. Inch it back and forth until they match. If the frame rate of the audio and the frame rate used for rendering do not match, the audio will begin to skew over time. Scaling the audio or the image sequence by multiplying it with the appropriate factor matches the frame rates but the sound quality or image quality will suffer. Avoid mismatched frame rates by making decisions about frame rates in preproduction. When you have synchronized images and audio, output the sequence as a movie file (e.g., an .avi and .mov file). Apply a compression method if you need to reduce the file size.

After going though your production pipeline, all the pieces will be together in your final product. It is time for the work to be seen by others.

4 Cleaning and Editing Data

The very first thing you need to do after capturing data is cleaning and editing the data. You want to do as little data cleaning and editing as possible. Good preproduction, especially a well-thought-out shot list, and the best calibration possible help you to generate good data. That reduces the need for data cleaning and editing. However, there is always some need for data cleaning and editing caused by the limitations of the capture system, space, and capture subjects' physical abilities, and other unpreventable reasons. So, you need to know some techniques for cleaning and editing data. In the first half of this chapter we will focus on cleaning and editing marker data; in the second half applying marker data to the skeleton. (Editing skeletal data will be covered in the Chapter 5.)

4.1 Cleaning Marker Data

In this section we're going to go in depth on cleaning motion data. We will spend a lot of time here because the concepts and practices in this section are very important to getting good motion.

4.1.1 Types of data

Different types of mocap systems generate different types of mocap data. It is important to know what kinds of problems the type of data that your system generates can have. Let's think about marker data that is generated by optical and magnetic systems and skeletal data.

4.1.1.1 Optical marker data (translational data)

Optical marker data generated by the optical mocap system is the simplest type of mocap data. It is translational data, that is, the information on the positions of markers that move in space and time. The cleanliness of optical marker data is heavily dependent on the cameras being calibrated well and all the markers being inside of the calibrated volume. The optical system does its best when it can see and track all the markers. When no camera can see a marker for a period of time and all or some of the cameras start seeing it again, there will be a gap in data where the marker is invisible to the cameras. The information about the marker's position is missing in the gap. The majority of cleaning optical marker data is filling gaps in data where the cameras could not see some markers. (We will look at a few methods for filling in gaps soon.)

4.1.1.2 Translational and rotational data

Data generated by the magnetic system is different from optical marker data: it has information on both rotation and translation of tracking sensors. The sensors can be followed when they are inside of an electromagnetic field generated by the transmitter source. Having the field properly calibrated for any potential issues inside the field, such metallic objects in the space that may distort the field, is the key to getting quality data. There are a few methods for cleaning this type of mocap data (rotational and translational). Because of the rotational information contained in the data, cleaning rotational and translational data is more complicated than cleaning translational data. However, data cleaning methods work fairly well on translational and rotational data if a capture subject stays inside the calibrated magnetic field.

4.1.1.3 Skeletal data

Skeletal data comes directly out of skeletal systems, most of which are based on optical mocap systems. Skeletal systems fit a skeleton inside a cloud of markers to generate joint rotation angles. Skeletal data has a full skeletal hierarchy of joints where the hips (root joint) have translational and rotational data and the other joints have rotational data only. (There are types of skeletal systems that allow the back and other body parts to have both translational and rotational data, in addition to hips. The best of these systems give you more flexibility than the ones that allow only the hips to have both translational and rotational data.)

If markers are incorrectly tracked, problems may arise in skeletal data. For instance, if a hip marker is not tracked properly, that can affect not just the hip motion but also the entire body motion. If a shoulder marker is missing, that can severely change the motion of the arm attached to the shoulder. When such problems happen, you have to either try to retrack the motion in post-processing or try to edit the motion, which is both difficult and time consuming. Pay attention to the outputs of skeletal systems and know what types of tools are available to deal with issues in data.

4.1.2 What to clean and what not?

What do we need (not) to clean? This is one of the hardest questions to answer. We have touched on this earlier, but let's go into more detail now.

4.1.2.1 What not to clean?

One of the obvious pieces of motion that won't be in final product is a T-pose. T-poses give an easy view of all the markers and help the system and you to identify markers. They are not in your shot list or storyboard. T-poses are often added to the beginning and ending of a shot to help with marker identification, but those are cut from the final motion. So, you don't need to clean T-poses.

If you are working on a motion that will be seen in a static camera shot, then you only need to clean the motion data for that camera shot. Only what the camera sees will be on the screen. So, there is no need for cleaning things that won't be seen. For instance, even if your camera moves and it keeps framing only the character's face and upper torso (i.e., you are using a dynamic close-up shot), then there is no reason why you need to clean data for the character's legs. Check what the camera sees using the aspect ratio that will be used to render the final product and determine what needs to be cleaned and what not.

Most mocap systems want to track and identify all of the data markers that they expect to see, no matter how many markers are actually in the mocap volume. The systems are usually designed to solve for a set of unknowns. If a system is unable to find some markers it will try to create them where it thinks they should be. It does so by taking very unreliable fragmented data and identifying it as missing markers. As the result, on the edge of the mocap space you tend to have a number of small segments of data. These usually cannot be cleaned.

Let's think about the motion data of a capture subject walking into a capture space, stopping in the middle, and then walking out of the other side of the space. When the subject is outside the space, the cameras cannot see the markers. As the subject approaches the space and starts to enter, various markers begin to appear to the cameras and are identified. Much of this data is compromised at first, but as the subject gets into the mocap space, it starts to solidify and become useful.

Because the data is very shaky at the very edge of the space where some markers are visible and others are not, you should discard this part of the data. You probably can identify the existing markers and recreate some data for some of the missing markers, but it is a difficult task since each marker usually moves at a different rate or in a slightly different direction than the other markers. Keep the frames that have all the markers and cut off the frames that have an incomplete number of markers but try not to cut off too much.

4.1.2.2 What to clean?

The best approach to data cleaning in general is to clean the data as much as possible at the marker data level. You need to try really hard to clean your marker data as best as you can if your client can't tell you exactly what will be in a shot (e.g., your client hasn't decided which camera position will be used for the shot). Cleaning marker data until it becomes the best it can be will create a good foundation for the motion no matter what your client decides to do with it.

If your motion will be used in a video game or other interactive application where the player/user has full control over the camera (i.e., your character can be seen from any camera position), then you have to clean everything as well.

Clean a motion using a camera position, move the camera, check the motion in the new camera view, clean the motion again if you find a new problem, move the camera again, check the motion again in the new camera view, … keep repeating that until the motion looks good from any angle.

It is tedious to clean data thoroughly. But having good data will give you better results down the road. If you decide to not clean the data well and rush through the process, you can get into trouble with faulty data later on, and often the best way to fix it will be going all the way back to the marker data and repeat the process all over again. Take your time and clean your data in the right way.

4.1.3 Labeling/identifying

All markers need labels or identifications associated with them. Labels are usually short character strings that allow you to know which marker belongs to what part of the body. Sometimes labels are numbers that are usually logically grouped for arms, legs, and other parts of the body. Either way mocap systems need to know which marker belongs to which body part and most systems rely on you to give them the information. Some systems automatically identify markers by themselves, but we are going to take the scenario of having a system that does not.

Why is labeling important? It is possible that an optical system does not see a marker for the entire duration of a motion due to occlusion. Also it is possible that an optical system mislabels a marker if the marker re-emerges after being occluded for a period of time. Labeling allows you to identify problems like these by letting you see each marker through time and space.

Labeling also allows you to make sure that all the markers that you need are in a scene without getting confused even if extra markers (or objects that appear to be extra markers) are in the scene. The optical mocap system always tries to identify a marker and goes to great lengths for that. It may assign a marker label to anything if it loses the original marker. For example, let's suppose that you have a few extra markers left out on a table (i.e., they are in your scene but not part of it). Because you can easily tell that they are not part of your scene, you leave them in. Your subject covers up one of the wrist markers. It's possible the mocap system decides that one of the extra markers on the table is the missing wrist marker and identifies it as such. You can see the problem easily if all the markers are labeled. But if markers are not labeled it is probably not so easy since all the markers look the same.

Labeling can help you improve the continuity of a marker's trajectory as well. If a marker is occluded, there will be a gap in your data. Marker labels let you know which marker has a gap in its data. Filling the gap improves the continuity of the marker's trajectory. Filling data gaps is a big part of cleaning data.

Another type of common problem that needs to be fixed by cleaning data is overlying data. Overlying data occurs when an optical mocap system decides that a marker exists in two places at once and gives the same marker label to two pieces of data segments that overlap in time. Let us explain how overlying data can happen by giving you an example. Suppose that there is a marker placed on a capture subject's waist. The subject bends down and the waist marker is occluded. The subject is crouched over and then starts to crawl forward. Now several different cameras can see the marker from different angles. A computational error can make the marker appear to be two markers in the same waist area and the system concludes that somehow the same marker exists in two places at once. The marker data ends up having two overlapping segments. One segment has good data. The other one has slightly skewed data. You have to decide which segment belongs to the real marker and which one to a ghost. Looking at how the marker position is before the occlusion starts and after the overlying data segments end gives you some basis for deciding which data segment should be kept and which should be removed.

One other common problem is marker swapping that occurs when two markers' labels are switched. Let's say you have a marker on the back of the head. You lean your head back to look up. The head and upper back markers come in close contact. After that moment the mocap system decides that the head marker's data is the back marker's and the back marker's data is the head marker's. You need to find out where this swap occurs and swap the labels back so that they are associated with the correct markers.

Correcting marker swapping requires either retracking a section of the data after correcting the labels or "cutting" the marker data where the swap occurs. Cutting divides your marker data into two segments. One segment contains the data before the cut and the other one after the cut. Now you need to change the marker identification of the segment after the cut to be the proper identification. In the case above, you need to cut both the head marker data and the back marker data where the swap happens and correct identifications for the segments after the swap.

4.1.4 Data cleaning methods

There are two major types of possible data cleaning. One is cleaning marker data that is translational data for optical systems and translational and rotational data for magnetic systems. The other is cleaning skeletal data that is largely rotational data with a small amount of translational data. There are some common methods of how to attack the data cleaning.

4.1.4.1 Eliminating gaps

The most common problem with optical mocap data is the gap caused by an absence of data or by removing an irregular peak that is a result of incorrect solution for computing marker locations. Gaps caused by the absence of data can occur for many different reasons, but occlusion is probably the most familiar one. One of the places where markers suffer from occlusion most frequently is the hands. A gap in the data looks like Figure 4.1. Gaps can be overcome in a few different ways.

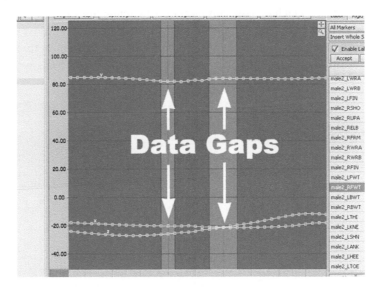

Figure 4.1 *Gaps in mocap data*

The quickest and most direct way to fill the gap is using linear interpolation (Figure 4.2). Linear interpolation is connecting the last good data point before the gap and the first good data point after the gap with a straight line between them. There is no ease in or ease out but a straight line filling the gap. So, the resulting motion may look mechanical for that section.

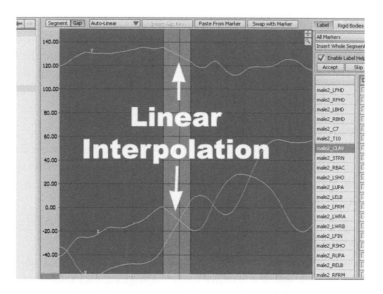

Figure 4.2 *Linear interpolation*

Another way to fill the gap is using spline interpolation (Figure 4.3). The type of spline most commonly implemented for a graphical data editing tool is a cubic spline defined by the positions and tangents of two end points. Spline tools normally match the tangent of the curve's beginning point to the tangent of the last data point before the gap and match the tangent of the curve's end point to the tangent of the first data point after the gap. Thus, the positions as well as the velocity

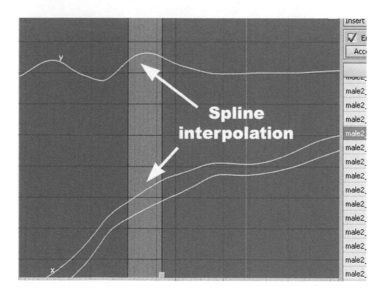

Figure 4.3 *Spline interpolation*

(the rate of change in position) are interpolated by the spline interpolation method while only the position is interpolated by the linear interpolation method.

Since spline interpolation uses more information from the data before and after the gap than the linear interpolation it provides a more organic curve to fill in the gap. Spline interpolation usually works well, but because of the nature of spline that interpolates the velocity as well as the position you may end up having a segment that has a lot of unwanted wild motion over a small area. Widening the gap by deleting small amounts of data before and after the gap and then applying spline interpolation will give you a smoother curve (i.e., a better result).

One other way to fill the gap is to add data by hand. It is not as tedious as you might think and this method can give good results. To do this, first you create a spline curve that fills a gap and then add as many control points as you need onto the newly created curve (Figure 4.4). The positions and tangents of the control points can be adjusted to create the motion that you want. This method gives you much more control over the shapes of the curves to fill in gaps than using splines with no additional control points.

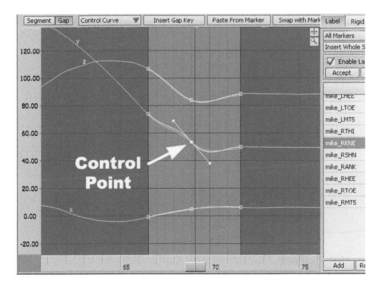

Figure 4.4 *Control point*

It's also possible to use a sample of similar data to fill the gap. Let's say you have a marker on your head that goes missing for a few seconds for some reason. It would be sensible to think that you could "borrow" the data from a different head marker and plug it in (Figure 4.5a). They are, after all, on the same body and therefore moving in the same direction and speed. The problem with this is that the markers may be rotating around an axis with an arbitrary orientation (Figure 4.5b). The rotation axis can be changing its orientation over time as well. This method can work OK at times, but you need to be very careful with using it because movements of markers are not so predictable. A better approach is the rigid body, which we will discuss in Section 4.1.4.3.

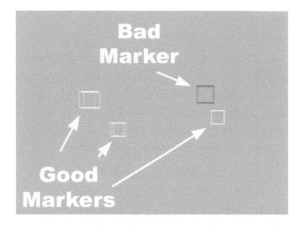

Figure 4.5a *Borrowing data*

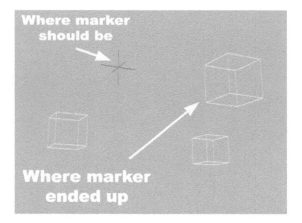

Figure 4.5b *Problem with borrowed data*

4.1.4.2 Eliminating spikes

Another issue that often crops up with any type of mocap data is what's called a "spike." The spike is a portion of data that suddenly leaves the normal range of the data (Figure 4.6). This results in skewing your data and causes a sudden jump in marker data that will eventually make your character look as if it suddenly jumped or twitched. Spikes are very easy to spot and also very easy to get rid of.

When you look at a graph of marker data a spike usually stands out as a sharp peak pointing up or down. Select a small segment around the peak and delete it (Figure 4.7a). Simple linear interpolation will usually fill in the resulting small gap fairly well (Figure 4.7b). You may decide to simply move the data point at the top of the spike instead of creating and filling in a gap. Either way, spikes can be fixed quickly.

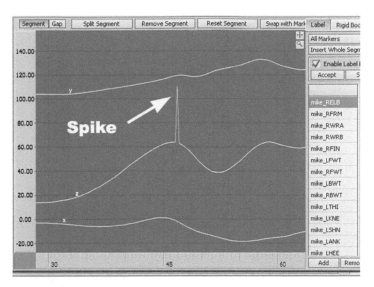

Figure 4.6 *Data spike*

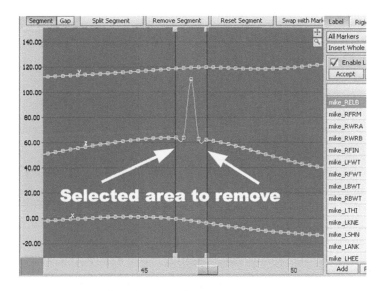

Figure 4.7a *Spike to be removed*

Helpful Hint: When removing bad marker data do not get sloppy or remove too much good data. However, removing a small amount of good data on either side of the bad data often yields a better result when you recreate deleted data.

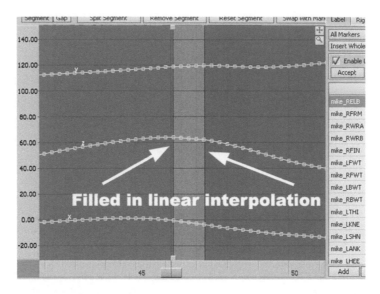

Figure 4.7b *Deleted spike filled in*

There are algorithmically more sophisticated methods that require more mathematical computations but often provide better results with less manual labor than the methods that we already looked at. Let us now take a look at those. First we'll look at the rigid body to fill a gap and then filters to get rid of peaks.

4.1.4.3 Rigid body

The rigid body is a good method to try when a data gap is wide and none of the methods described in the previous section works well.

A good place to use a rigid body is a section of a body that is rigid (e.g., the head) or relatively rigid (e.g., the rib cage and hips). It is because when a "rigid" body section moves, the markers on the section move with it without changing their relative position among them. On the other hand, when a flexible part of a body moves, for example an arm, the relative position among the markers on the arm changes as the arm moves. A hand is an OK place to use a rigid body because the metatarsals (bones in a hand) can move independently (but not much).

Creating a rigid body is like creating a polygon using markers. You need at least three markers to create a polygon (Figure 4.8a). If you have only two markers (Figure 4.8b), then it won't work well since the polygon can rotate freely around the line defined by the two markers. One marker (Figure 4.8c) does not define any orientation at all.

The markers that you select define a rigid body and the rigid body's motion is derived from the markers. When one of the markers in a rigid body is hidden, missing data will be created using the position and orientation of the rigid body and the position of the marker with regard to the rigid body before it disappears.

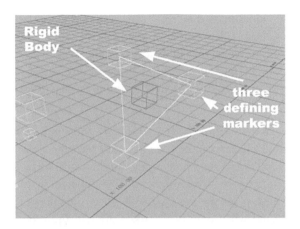

Figure 4.8a *Three markers*

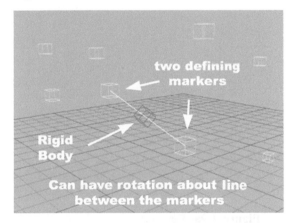

Figure 4.8b *Two markers*

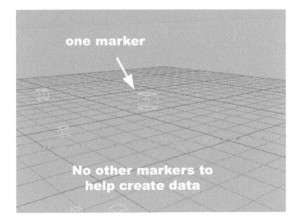

Figure 4.8c *One marker*

To use this method, in addition to the "bad" marker(s) that has a data gap, select at least three "good" markers with no gap. (The number of good markers can be four, five, six, or more.) You need to make the selection in a section of data where the "bad" marker is not bad (i.e., it is not missing data). All markers, both good and bad markers, need to be rigidly bound to each other. In other words, they don't flex, bend, or move in relationship to each other. Create a rigid body using the selected markers.

If not enough markers are selected to create a rigid body, there will be some problems. An example is a rigid body of four markers on the head (Figure 4.9a) where two of them have a gap to fill in. Let's suppose that you create a rigid body using four head markers when the subject is standing in the middle of the mocap space looking forward. The subject now turns around and two of the head markers are occluded. If you use the rigid body the result will not look right for your motion (Figure 4.9b). It is because when only two markers are left, the rigid body doesn't have a third

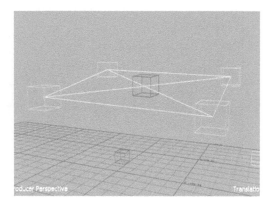

Figure 4.9a *Rigid body*

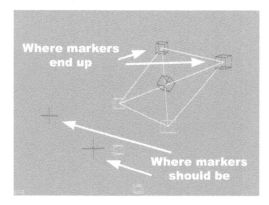

Figure 4.9b *Rigid body with bad marker placement*

marker to orient itself and the rigid body can spin around the line defined by the two markers that are still visible.

One last word of caution: When creating a rigid body, make sure that the markers are all present and at correct locations. If a capture subject starts outside the capture volume and then comes into it, the markers at the beginning are all over the place and it is not a good choice to set up a rigid body there. A better choice is when all the markers are visible and obviously in a correct relationship to each other. The software is usually not smart enough to know where the markers or rigid bodies should be, so use your best judgment.

4.1.4.4 Filters

The filter is a mathematical application that smoothes data by reducing low- or high-frequency noises, detects and removes sudden changes in data, or changes other properties of the data. There are a few different types of filtering algorithms for mocap data.

Why do you need filters in the first place? Why is motion jittery and almost never 100% dead on and correct? The answer to the second question usually lies in the accuracy of the hardware (physical equipment) that generates raw data and the mathematical algorithms implemented for the software that processes the raw data into a more useful one. There is always an inherent amount of error in the hardware and software. The error gets into data and becomes evident in the motion. From afar, the motion will look very accurate, but if you get close to the markers you may see them jiggling and moving slightly. Even with the best calibration this can occur. Low-pass filters remove jittering by not allowing high-frequency noises to be passed into filtered data. That's one reason why we may need to run filters. Another reason is spikes in data. When a marker is invisible due to occlusion or some other cause the mocap system can make a wrong guess and create a spike in data. Low-pass filters can remove spikes (Figure 4.10a and Figure 4.10b) much more quickly than the manual method we talked about

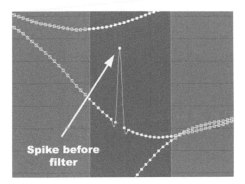

Figure 4.10a *Data spike*

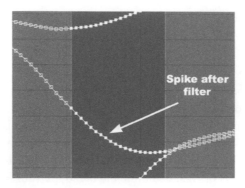

Figure 4.10b *Filtered data spike*

in Section 4.1.4.2. When a calibration is poorly done, high-frequency noises tend to be compounded. An excellent example is the Butterworth filter, which does a remarkably good job of smoothing out spikes without smoothing the rest of the data too. Low-pass filters remove high-frequency noises (Figure 4.11a and b).

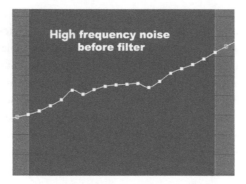

Figure 4.11a *High-frequency noise*

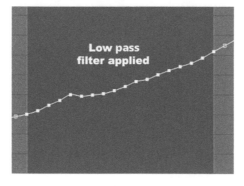

Figure 4.11b *High-frequency noise filtered*

Each filter has its own function and settings that allow you to change the filtering conditions. Each generates a different result even if it's applied to the same data. There are several filters that are implemented to perform multiple tasks, for example, remove both low- and high-frequency noises. You need to try different filters and decide which one solves your problem with the best result.

Be aware that too much filtering can deteriorate your data, instead of improving it. For instance, smoothing filters work well to make your data look more stable if there is a lot of marker shake. But also it can make your resulting motion look too smooth. Look at the quicktime or .avi file on the CD. Notice the foot twisting with an over-exaggeration of smoothing.

One of the major reasons why we capture motion is to get the subtle characteristics of balance, poise, and motion, and apply it to something later. A large part of what makes our motion look "human" or at least alive is in how our muscles react to move our body. It's not in a linear curve or even necessarily a smooth curve all the time.

Once you start smoothing data, you change all the velocities and accelerations, as well as positions. If you don't stop where you should, filtering can make the motion appear no longer real. Although at times it's critical to get rid of the jitter that's usually inherent in most mocap systems, you need to be judicious with smoothing. Try to use the absolute minimum filtering that you can. Never overdo it or the results will make your final product look "floaty" and strange as if your character were skating through syrup.

Some people will run filters only on the upper body markers and not on the lower body in order to make sure the feet stay in the best position.

4.1.5 When to stop?

This is a good question and one that you will have to figure out yourself. It really depends on what level of cleaning you're looking for. Let's look at a few different scenarios.

When you only want to see your motion for either shot blocking or to get an idea about how the shot will look, don't worry too much about filtering marker data. This is not your final motion, so you don't need to fix everything at this point. In this case, it is a good idea to apply the marker data that is not fully cleaned to a skeleton, look at the motion on the skeleton, and find where problems pop up. Go back to the marker data and fix the parts of the data with the problems. If there are some problems that cannot be solved by working on the marker data, then you need to edit skeletal data. Going back and forth between marker data and skeletal data can help you decide what to fix, where to fix, and when to stop.

When you want good motion because you need for someone else to see it, and you don't want him/her to think you slapped it together, it's best to run a few filters across it and make sure there's no marker swapping or mislabeling. You probably need to do this on a shot by shot basis. Do not kill yourself over little things. However, if anything immediately pops out to you, it's not a little thing; it's a big enough problem that it stands out. Whenever anyone asks us "Is this a problem?" We say "If you've noticed it, it's probably a problem." The feet and hands are usually the places that are the most prone to small errors. Make sure the feet are stable and the hands are not slightly shaking. You may have to remove some data and replace it for both of these examples.

You can expect the most tedious data cleaning task when you want to perfect your mocap data and say "I cannot do anything else to it." Remember that just like key-frame animation you could spend all your time tweaking and changing the data. You could burn yourself out or drive yourself crazy. Give yourself an acceptable standard, and don't go above that unless you have extra time, which you probably never have.

Get someone else to look at your motion critically and ask if it's good enough or if he/she sees any problem. Remember that you are dealing with mocap data of human motion most of the time and that all humans are experts at observing human motion. We are much more familiar with human motion than motion of animals, for example dogs and cats. However, it is often difficult for us to pinpoint what is exactly wrong when there is something that looks odd. Try to find someone who can spend time examining your motion and tell you what she/he thinks, instead of just saying "Yeah, it's great." You want objective opinions of other people.

Even after you have moved to the next stage (or a much later stage) in your pipeline, do not be afraid to go back to the marker data and make changes on it. Remember that with optical data there is always the 2D camera data to go back to as well. So, no matter where you have gone to in your pipeline, you can always start over from the very beginning if necessary. If you are using a system that does not allow you to go back to earlier forms of the data, always save original data into a file, make a copy of it, and edit the copy.

4.2 Applying Marker Data to the Skeleton

In this section we will go over the process that takes your cleaned marker data and attach it to a skeleton that you have created. What you will get at the end of the process is a skeleton with a motion driven by mocap data. Since clean marker data is essential for good motion, marker data should be as clean as it can be. We are going to use a pipeline that is designed with Maya and MotionBuilder (Appendix B), but the process should be similar to that in a pipeline with other software packages.

To apply marker data to a skeleton in MotionBuilder we will go through a couple of steps. In the first step we set up what MotionBuilder calls the "Actor" and in the second step we set up what MotionBuilder calls the "Character."

First we set up an Actor which is a set of constraints represented by body parts that can be rotated and translated. Rotating and translating the Actor's body parts allow markers to be on desirable parts of the Actor. (The Actor's body parts can be scaled, but scaling is not part of the solver.) We can tweak the relationships between the markers and the Actor so that the markers drive the Actor's motion in the most desirable way.

Secondly we set up a Character that is a list of relationships and joint names. When a skeleton is attached to the Character, the Actor feeds its motion into the proper joints of the skeleton by way of the Character. We can adjust the relationships between the Character and Actor so that we can get the best movements in every part of the skeleton.

You might ask "Can we forget about the Actor and the Character and just attach the marker data directly to the skeleton?" or "Are there options to do this with other programs?" The answer to both of these questions is "Yes." There are other applications that use different methods, such as the Giant Studios tools, Diva, and the PEEL solver. Even in MotionBuilder there is the ability to take marker data and directly apply it to a skeleton, usually using a rigid body constraint, but this is normally reserved for props or joints in addition to the primary skeletal joints.

The Actor and the Character can be considered as a pair of mediators whose handshake has to occur in order for information exchange to happen. One mediator works on the marker data side and the other one on the skeletal data side. When the handshake between the two mediators is performed properly, the information is successfully transferred from the marker data to the skeleton.

The Actor and the Character are probably the least intuitive (hence, the most confusing) part of MotionBuilder. However, MotionBuilder is equipped with a number of good tools for skeletal data editing and is widely used for that purpose.

Let's begin with setting up the Actor. We will try to explain it step by step.

4.2.1 Actor

You want to start in MotionBuilder with markers that are in a T-pose. You usually have some sort of calibration motion or throwaway motion from the shoot that begins or ends in a T-pose. Import a file with markers in a T-pose into MotionBuilder. You should see markers as a collection of dots or point cloud (Figure 4.12).

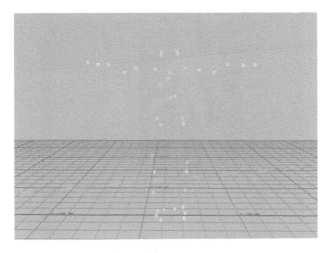

Figure 4.12 *Markers in T-pose*

Now you need an Actor. The Actor is a collection of constraints and it is important to manipulate it to a degree that it matches as close to your marker data as possible. To import an Actor, go to the Asset Browser, click on the Character folder in the Templates folder, and drag the "Actor" icon onto the 3D viewer. Move the Actor over to marker data that's your T-pose and move it around until the hip markers fit inside of the Actor's hips (Figure 4.13). Using the translate manipulator (T key) on the hips lets you move the entire Actor around.

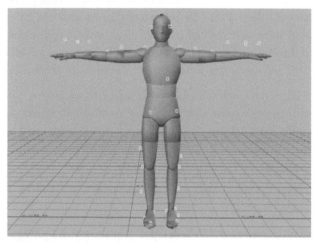

Figure 4.13 *Actor over markers*

Next you need to scale the overall Actor to get it close to the size of your markers. When needing to scale everything, you can select all of the circles in the "Actor Controls" box (Figure 4.14) and

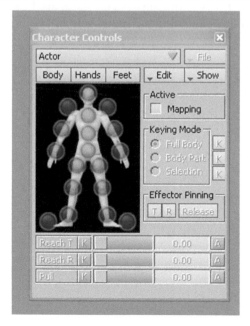

Figure 4.14 *Actor controls*

select the scale function by either clicking the scale icon or tapping the "S key." Now rotate and scale the different body parts of the Actors to get the best possible match to the point cloud (Figure 4.15). The upper arms often need to be rotated to better match the hands. The legs and feet also often need to be rotated.

Remember that MotionBuilder is generic 3D animation software. It has no idea what type of data you are using or how many markers you are using. Some of the markers that you captured in your mocap software may not be needed in MotionBuilder although your mocap software may have used these markers to solve some problems on its end. Sometimes these extra markers can be used in other ways to help with the overall motion of the character. An example is a marker in the middle of the lower arm. It may not be used for the Actor setup but can be used for other purposes, such as controlling the roll (twist) of the lower arm along the longitude axis of the lower arm.

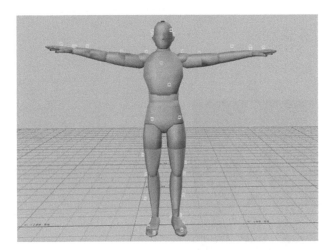

Figure 4.15 *Matching Actor to markers*

Now that you have the Actor and the markers lined up, it's time to associate the markers with the body sections of the Actor. Click on the Actor under the Actors node in the Navigator panel. The Actor Settings shows up with a gray figure and empty boxes. To the right of the figure is a button labeled "MarkerSet ..." (Figure 4.16). Click on the button and choose "Create." You will now see several small circles appear around the gray figure.

When the body sections of the Actor are ready to be associated with the markers, select markers for a body section and Alt-drag them into the appropriate section of the gray figure. For example, select markers for the head and Alt-drag them into the head of the gray figure (Figure 4.17). (Since

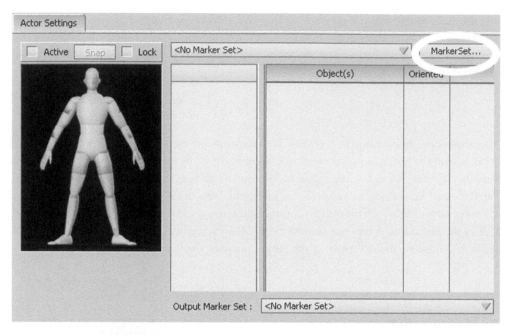

Figure 4.16 *MarkerSet button*

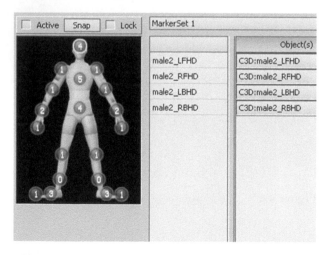

Figure 4.17 *Associating markers with body sections*

MotionBuilder has a fairly comprehensive overview of this process in its documentation we are not going into every detail here.)

Once all the body sections of the Actor are associated with appropriate markers "activate" the Actor by checking the "Active" box in the Actor Settings. You can play the animation and see the

Actor move with your marker data. A range of motion are the best to test this setup. If markers have been associated with a wrong body section, you will see the Actor moves in an obviously weird way. Mistakes tend to happen in the hands and feet segments. When you find a marker (markers) associated with a wrong body segment of the Actor, move it to a different body segment. Even after you finished assigning the markers, you can still tweak the positions of the markers by translating and rotating the Actor's body segments. (See .fbx file of markers and Actor on CD.)

4.2.2 Skeleton

Now that your Actor is moving, it's time to attach your skeleton to a Character and "characterize" it as MotionBuilder calls the process. But before we start explaining how to set up a Character let us give you some hints on how to build a skeleton that works well in MotionBuilder. No matter which 3D software you use to create a skeleton, there are some rules that seem to make MotionBuilder happier. In this section we will be telling you a few hints largely based on those implicit rules that we have discovered. (A new version of MotionBuilder will probably be out by the time you read this book. The new version may have a different set of preferences. So, as with everything in this book, use the following as a guideline but not as a be all end all.)

Hint 1. When you are going to build a skeleton, marker data in a T-pose can be brought into your 3D software package and used as a guide for the proportion of your skeleton.

If your 3D character can have the proportion of the marker data (i.e., your capture subject), build the character's skeleton using the capture subject's proportion. In that way, there will be less need for skeletal data editing, especially retargeting limbs. However, being able to use a capture subject's proportion for a 3D character is rare. You are more likely to be building a skeleton for a 3D character that's been modeled or based on a model sheet of a 3D character that's been drawn but not modeled. Either way the skeleton of the 3D character has to be proportioned to the 3D character, not to the capture subject. A MEL script "marker_lookat.mel" which generates spheres at the locations of markers is on the book's CD. If you want to build a skeleton using marker data as a guide for the skeleton's proportion in Maya, the script visualizes your marker data for you.

If possible, cast the talent whose size and proportion matches that of the 3D character as closely as it can. Men and women are proportionally different, so are children and adults. Capture a woman, not a man, if you need motion for a female character. Capture a child, not an adult, if you need motion for a young character. The closer the match between the markers (your capture subject) and the skeleton (your 3D character), the more smoothly everything will work out. Avoid forcing MotionBuilder or Maya to compensate too much for differences in sizes and proportions. The more that the software has to change data, the more likely problems will creep into your pipeline.

Hint 2. Add an empty node named "reference" at the top of the skeletal node.

Having a reference node helps because there tends to be a 10 to 1 scale difference between MotionBuilder and Maya. You can apply scaling to the reference node when the entire skeleton needs to be scaled up or down. The .fbx export in Maya has a unit conversion setting, so you can also address the issue there, but it's always a good idea to have the "reference" node.

Hint 3. Build a skeleton as you want to, but remember that MotionBuilder will want it to be rotated into a T-pose.

You can build skeletons in any fashion; however, remember that MotionBuilder will eventually need them in a T-pose. Have your skeleton's rotations "zeroed" out in Maya so that there will be no rotational values left in the joint rotation channels when the skeleton is imported into MotionBuilder. Once you have your skeleton in MotionBuilder, if it is not in a T-pose, put it into a T-pose by rotating joints. Make sure to do this before characterizing it. (Also, before the characterization, make sure that your skeleton is looking down the global Z-axis.) If your character (skeleton) is not in a T-pose before the characterization, you'll have to use reach constraints to make all the body positions match. That will create an unstable solution.

Hint 4. Orient all skeleton joints so that their local rotation axes become identical to the global space's axes (Figure 4.18).

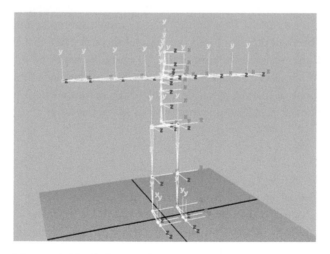

Figure 4.18 *Skeleton without rotations at joints*

If you leave rotational values on the skeleton joints, you may have a problem especially when trying to blend motions. The end result after the blend may look fine, but when you are in the middle of using MotionBuilder's blend tool, the skeleton may look either skewed or as if it had been rotated

90 degrees about the hips. Although you really need to see how blending is working during the blend process, this makes it almost impossible for you to see the result of the blend until the process is finished and rendered (more about blending will be covered in Chapter 5). In order to get all rotational values out of the joints in Maya, apply "Freeze Transformations" (with every option turned on) to your skeleton's joints. It should be noted that this only applies to the Blend tool. The Story tool has been added to MotionBuilder and supported more over the years. It is far more flexible than the Blend tool. We'd suggest that you go through the tutorials on both Blend and Story tools and see which will work best for you.

Hint 5. Name your skeleton's joints using MotionBuilder's names for joints.

Name your skeleton's joints in the way MotionBuilder wants them to be named. (Please reference MotionBuilder's Character for these names.) Using their naming scheme makes things go quicker although this is not something that must be done.

When you finish building your skeleton, export it out as an .fbx file. You will import the file in the next section. (Sample Maya and .fbx files are included on the CD.)

4.2.3 Character

It is time to attach a skeleton to a Character. First, use the Merge option under the File to import the .fbx file that contains your skeleton into MotionBuilder. The skeleton should show up in the middle of the space. Go to the Asset Browser and look in the Character folder in the Templates folder and drag a "Character" into the 3D View window. If you like, you can drag it over your skeleton's hip and it will try to automatically characterize the skeleton, but this will only work if everything is named as MotionBuilder expects it.

> If you notice that your skeleton is extremely small once it's imported from Maya into MotionBuilder, then select the "reference" node. Go to the "Properties" window and change all the scale sizes to 10. You should now have a skeleton that seems to fit in the world. Remember that Maya .fbx export allows for unit conversion or you can scale the skeleton up in Maya before exporting. Try a few different ways to see what works the best in your pipeline.

If you don't characterize it automatically, you'll need to drop the skeleton joints into the appropriate slots in the Character Definition tab. If you have everything named with the names MotionBuilder is looking for, you can just Alt-drag them all down. If you didn't follow the naming convention, you'll need to Alt-drag each joint one at a time and associate it with the proper section of the character.

Once that's done, check the "Characterize" box. You'll now see a couple of questions such as if you want a quadruped or a biped. You may even get a few errors. If MotionBuilder lets you go on, it's probably OK to ignore those messages for now.

In the Character pane, change over to the Character Settings and choose the Actor as your input type and then choose the Actor that has marker data driving it as the Input Source. Click the Active box (Figure 4.19). When the Character is activated, you see your skeleton moving around with the actor (Figure 4.20), but it may not be in the same place, and there will more than likely be some other discrepancies. We'll cover such issues in skeletal editing in Chapter 5.

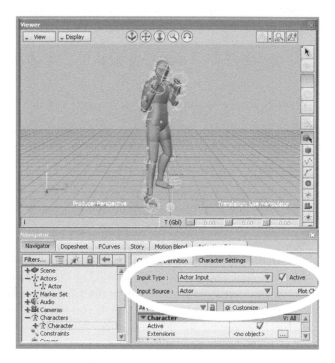

Figure 4.19 *Activating Character*

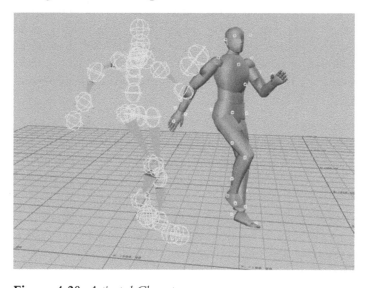

Figure 4.20 *Activated Character*

Apply some motion and take it back into your 3D package to see if it looks OK and imports correctly. In order to make sure that you won't have to redo it all over after doing a lot of work, look at the skeleton with some motion applied onto it in the 3D package.

To get a motion into your 3D animation package, you must "plot" the motion to the skeleton in MotionBuilder (Figure 4.21a). Plotting takes all the motion that's currently just pointed at the skeleton and bakes the translations and rotations in every frame as a key-frame. The initial dialog will ask if you want the motion plotted on the skeleton or the Control Rig. Choose the skeleton (Figure 4.21b). Then you see another dialog box with options. The default setting usually works fine. So, just choose to plot (Figure 4.21c). Once the motion is plotted to your skeleton, select all of your skeleton's joints and export the skeleton as an .fbx (animation only) file. Exporting and "Saving As" are two different things in MotionBuilder. Do both but only import the exported file into your 3D animation package.

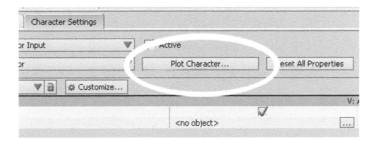

Figure 4.21a *"Plot" the character*

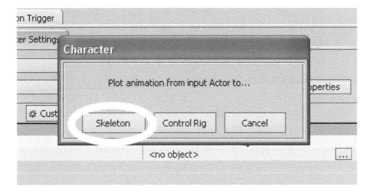

Figure 4.21b *Choose skeleton*

Go to your 3D animation package and first open the file that you exported the skeleton from. In addition to the original skeleton with no motion, the file may contain the skin geometry bound to the skeleton, 3D models of the environment, textures, lights, and anything else needed for the scene. Now import the .fbx file that contains a copy of the skeleton with motion plotted

Figure 4.21c *Use defaults*

to it. The merge option tends to work the best, but different instances may call for you to play around with the options. Once imported, the plotted motion should be dropped onto the correct joints of the original skeleton and you can see your 3D character moving. You can repeat this for every shot.

Just as a note, there is no reason to just import the skeleton without skin into MotionBuilder. We did so to give you a better look at the skeleton itself and the marker to Actor relationship. Whenever possible, export your skeleton with skin bound to it out of your 3D animation package. By importing the skin bound skeleton into MotionBuilder you can see if the character's arms intersect with the torso, how far the bottom of a foot is to the ground, and other things that you can check only with the skin bound skeleton. If you find, for instance, a character's hand keeps going through its thigh, you may have to play around with the marker to Actor relationship or add in an offset to counter that.

Another note is that MotionBuilder at times has trouble with some of the tweaks that are applied to skin geometries in Maya. Applying "Delete All By Type/Non-Deformer History" to your skin gets rid of those instances. Also there have been issues with cluster deformers in the past. If you run into a number of problems with a skin bound skeleton that is imported into MotionBuilder, we suggest that you create a low-resolution model that has just enough definition for edges of hands, feet, etc. and use it, instead of the real skin geometry, while editing data in MotionBuilder.

In next chapter we will look at skeletal editing and blending in detail.

5 Skeletal Editing

This chapter is about various ways to edit skeletal data, that is, data that's rotational in nature. The skeletal editing techniques that will be discussed include retargeting, blending, inverse kinematics (IK), floor contact, rigid body, looping, and poses. Let's start with retargeting.

5.1 Retargeting

When we have mocap data that has been applied to a target skeleton, it is very possible that the source skeleton (the skeleton in the data) and the target skeleton do not have the same proportion. That is almost always the case between the capture subject whom markers were attached to and the 3D character that you want the motion to go onto. Some allowances must be made for the proportional differences in order for the motion to fit the target skeleton as well as possible.

Retargeting is more than just slapping the rotations and translations from one skeleton to another. It is about trying to adjust for the proportional differences while keeping the motion from suffering, looking too stiff, quirky, or weird. Retargeting itself is a discipline and one of those challenging research areas that people are constantly working on.

The key to retargeting differs depending on the type of motion you are working on. A retargeting strategy works well for a walking character but may not work for a character crawling on the ground. The retargeting setup for a quadruped or non-biped creature must be different from the ones for biped humans. To deal with a variety of situations, retargeting requires sets of strategies. We will discuss a retargeting method for a biped human here. We would like you to look at the retargeting process as a whole and what it does, and understand why you do what you do. That way you can develop an effective retargeting strategy when you come to a scenario that we do not cover in this book.

5.1.1 Reducing need for retargeting

The classic retargeting problem is "We'd like to capture this actor who is 6 feet and 2 inches tall but apply the data to the character that will be only 3 feet tall." Even with casting, it is extremely difficult to find a 3-foot tall person for the motion (Figure 5.1). Kids do not have the same proportions

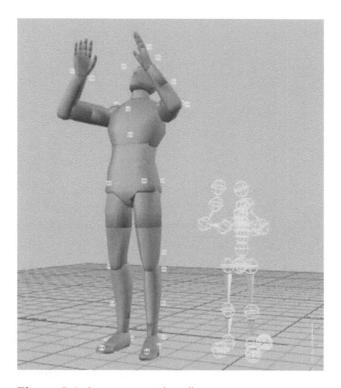

Figure 5.1 *Large source and small target*

or move the same way as adults. Look at a book on figure drawings. You can see differences in proportions between adults and children, men and women. Drawing in a sketchbook can help you create differently proportioned skeletons, instead of trying to use one skeleton based on an adult male for all of your characters.

You need to work closely with whoever is directing the mocap shoot as well as anyone else who has a say in the final outcome. Should the 3-foot tall person act like an adult, a child, or something altogether different? You will be editing motion differently for each scenario. Understand what is needed as well as what can be done during the mocap shoot to solve possible problems before they happen.

One method is to have a scaled-up set in which everything is larger than normal. In the set the 6 foot 2 inch tall actor should behave like a 3-foot tall person. This method works well unless you have to mix several differently proportioned characters. Another method is to place multiple targets for differently sized characters in the mocap space. If a small character needs to reach up to hit a target (e.g., it is going to shake hands with a taller character), it reaches a higher target while the taller character reaches down to touch a lower target. This will get the captured motion closer to its target but it will still require editing.

5.1.2 Scaling a skeleton

If you are going to apply the motion of a tall person (the source skeleton) to a much shorter character (the target skeleton), what are the obvious problems? One is that a normal stride for the tall person is much greater than the smaller person. To match the stride, first try to scale the source skeleton down to make as close a match between the two skeletons as possible (Figure 5.2). Assuming that the feet are normally in contact with the floor, the length from the waist of the skeleton to the bottom of the foot is important. This is what you need to try to match up. Do not worry if the overall heights do not match up. In this case the waist to foot length is more important.

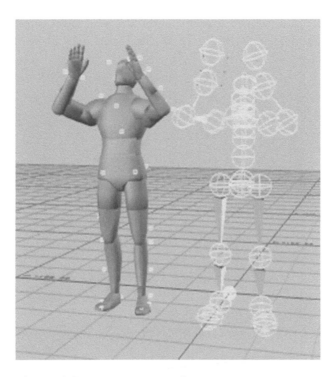

Figure 5.2 *Same size source and target*

You can scale the target skeleton up or the source skeleton down. Each will give you the same basic results. When an end character's size and proportion are set, you need to scale the source to match the character rather than the other way around. Either way write down your scaling factors so that you will use them consistently.

If multiple capture subjects who interact with each other are captured together, retargeting must be done without scaling source skeletons or with scaling them all using the same scaling factor. If you scale the source skeletons using various scaling factors, the target skeletons' relationship to one another will be skewed and none of their interactions will line up. Shaking hands is one of the best examples. If you capture two people who are 6-feet tall shaking hands and apply the data to a 3-foot tall character and an 8-foot tall character, the hands will no longer line up. This is a problem that you can overcome with retargeting to some degree. But think about it before the capture. Instead of relying on retargeting, have the actors compensate for the size differences during capture. Preventing or reducing the need for retargeting is important.

If the corresponding joints of the source and target skeletons are in different locations, you may see some unusual hyperextensions or the knees locking out, appearing very stiff (Figure 5.3a). If this happens, start playing around with the scale factor until the legs look more normal (Figure 5.3b). If you are really stuck and cannot get it to work well, try letting the target skeleton be a little larger than the source skeleton. The reason for this is that the target will always be able to reach at least what the source reaches. If the target is smaller, you are more likely to see the joints locked out or hyperextend.

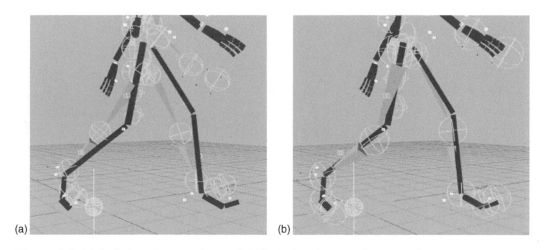

(a)　　　　　　　　　　　　　　　　　　　　(b)

Figure 5.3 *(a) Locked out knee rotation; and (b) Locked out knee rotation corrected*

5.1.3 Fixing foot sliding

Let's suppose that your source and target skeletons are pretty close and you have them matched up fairly well. There is no hyperextension or joint locking but the feet are still sliding a lot. MotionBuilder and other motion editing software allow you to use a type of IK that will adjust the target skeleton to place its ankles, balls of the feet, or toes exactly where that of the source skeleton is located (Figure 5.4a). Do this with both position and orientation of the ankles. If you apply this to the ankle and toe of the same foot simultaneously, beware that the constraints may fight each other. The foot may end up with even more sliding if the ankle and toe are constantly trying to push or pull the joints to match (Figure 5.4b). This is difficult to see when looking at a frame, but in motion you can see if the foot is unstable and trying to satisfy multiple constraints.

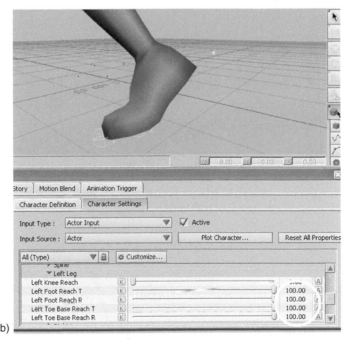

Figure 5.4 *(a) Ankle reaching source's ankle position; and (b) Toe and ankle reaching source's positions*

When working with the feet, turn off any type of influence to the hips. Many people leave it on at times but it is usually better not to mess around with the hip rotations and translations at all. The reason for this is that the hips are the root of the skeleton and normally the only source of translation. If you rotate or translate the hips everything else in the skeleton is affected. That can change the characteristics of the whole motion. Both the hips affected during retargeting and over-smoothed data results in a motion that looks as if it had no weight at all (called "floating motion"). This is one of the biggest criticisms on mocap, so try to avoid it.

5.1.4 Working on the spine

The spine is a difficult area to edit. Minor tweaks to it can cause large changes but sometimes it has to be edited. You want the spines of the source and target skeletons to match up fairly well but this is not always possible. Sometimes the source and target skeletons have spines with different lengths. If the source skeleton can bend over and reach the ground with its arms, the target skeleton, with different proportions, might not be able to reach the ground when it bends over. In Figure 5.5, you can see that the target skeleton in white has a shorter spine than the source skeleton in dark gray. The target skeleton's hand does not reach the ground whereas the source skeleton's hand does. This is where you may have to use some key-frames for your retargeting and also do what we earlier advised against, allow the hips to move. For every so-called rule in mocap, there are usually a couple of situations where you have to break it.

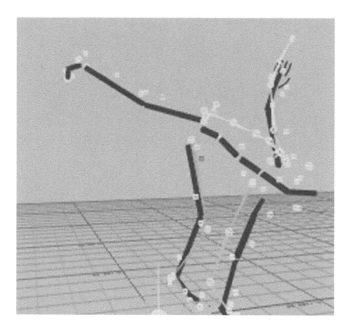

Figure 5.5 *Source reaching the ground but not target*

In this situation, you would want the hips to dip lower so the spine and arms can get closer to the ground. It is possible that you are using IK retargeting on the spine and hands. If the hand is supposed to touch the ground or an object, it is important that hands of the source and target skeletons are in the same location. This means loosening everything up a little. That will allow the target's hand to reach the ground. Most systems allow you to key-frame these changes on and off so that it does not affect the entire motion. You may have to have one setup just for this action. Be ready to have special setups for not so special cases like this.

If the back is also loosened up carelessly, it may look unnatural. If the back is not loosened at all, the shoulder may do some strange rotations in order to get the hand where it needs to go. Play around with these to get the best fit. MotionBuilder has stiffness and pull modifiers that can be used to change how much different parts of the body react. Usually adjusting the back (or torso), neck, and head gets the best possible match between the spines of the source skeleton (Actor) and target skeleton (Character). If these are not adjusted, it may appear that the back moves as one piece and does not curve or bend correctly. That will be very evident when the motion is put on the target skeleton, brought into your 3D package, and the motion is rendered out.

Retargeting is one of the most difficult things to do in skeletal editing. Before going into it, be aware that you will be required to do some tweaking. Being precise will save you from spending a lot of time on tweaking. In the next section you will learn how to blend two different motions together.

5.2 Blending Motions

Motion blending is one of the hardest concepts to learn, so this section may get a little long. Motion blending is taking two motions that have been applied to the same skeleton and merging them (Figure 5.6). If you try to blend motions on different skeletons, there will be a large number of issues that need to be taken care of. So, we will talk about blending two motions that are on the identical skeletons. It should be noted that this is all in relation to MotionBuilder's Blend tool and that the Story tool may be better suited to your application.

First make sure that two motions that will be blended are on identical skeletons. The two skeletons should have the same hierarchies and proportions. The corresponding joints in the two skeletons should have the same names and the same orientation of the local rotation axes. The lengths and orientations of the corresponding bones in the two skeletons should be the same. Blending is basically taking the rotations of the corresponding joints in a pair of skeletons and interpolating them. Different bone lengths could cause the legs not to reach the ground or go through it. Different joint orientations could cause unwanted rotations. The two motions to be blended should be on identical skeletons but not on two different variations of a skeleton.

If you are trying to make a loop or a very long continuous motion, you may use the same motion twice or more. Reasons to create the long motion by looping are numerous. If your mocap space is 20 ft × 20 ft, the most distance you can get is someone walking across the diagonal line in the space, approximately 28 feet. Let's say you need your character to walk the entire length of a football field, which is about 360 feet including the end zones. You will need to blend a 28-foot long

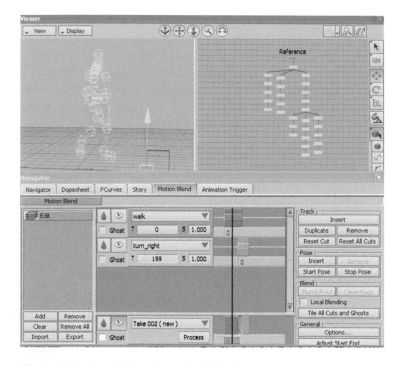

Figure 5.6 *One skeleton for multiple blend motions*

walking section to itself several times over to get it to cover the entire distance. (Looping will be more detailed in Section 5.6.)

5.2.1 Selecting a blending point

What will be blended are similar body positions and movement qualities from one motion to the other. This is where blending can get confusing. Think of each motion as a section of time. You have two sections of time that you want to put together. If you put them over each other, then you have a blend area that is too long (Figure 5.7a). If you just place the beginning of one next the end of the other, you have no area to blend across (Figure 5.7b). You want to slide one underneath the other, just taking up enough room for a smooth transition from one motion to the other (Figure 5.7c).

To demonstrate this, let's think about blending a walking motion (motion A) and a turning motion (motion B). Motion A will be a straight walk across the mocap space. Motion B will be a straight walk that then has a 90-degree turn to the left and then exits the space as a straight walk again (.fbx files are on the CD).

You need to find a blending point, that is, places in the two motions that are similar. It is important to plan for blending and give proper instructions to the capture subject. Otherwise your capture subject may walk at different speeds or with different body postures in the motions that you will need to blend. Have the capture subject walk in a relaxed manner at normal speed if that suits

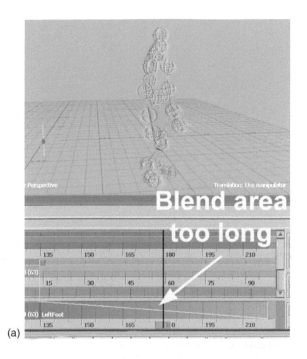

(a)

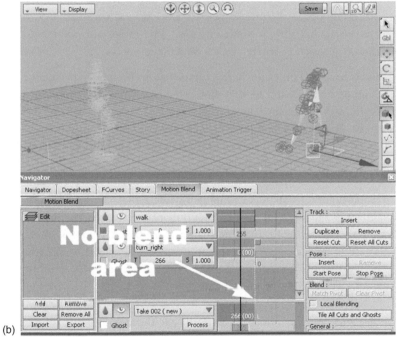

(b)

Figure 5.7 *(a) Blend area that is too long; (b) No blend area at all; and (c) Good blend area*

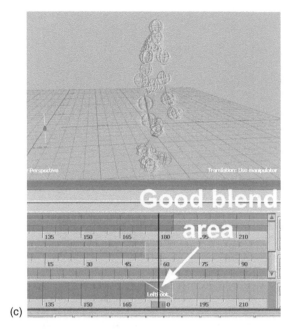

(c)

Figure 5.7 *Continued*

your project. Depending on the project you are working on, you may need to have the capture subject walk in a peculiar way or constantly run. The important point is to have consistency. Consistency makes it much easier for you to select blend points.

Suppose that you have one motion where a capture subject is walking at a fast pace and the upper body is bent slightly forward, and another motion where the capture subject is walking at a relaxed pace and the shoulders are slumped down. Thus, you have two very different paces and poses. Most blending software does not know how to change speed or posture in an intelligent manner, that is, it cannot correct inconsistencies between two motions. It simply lets a blended skeleton change from one pose to another in an unnatural way.

A common blend point for walking or running is when a character has one foot on the floor and is shifting weight across the foot. If the left foot is firmly planted on the ground, select the point where the right foot is passing the left leg as the mid-point of the blend (Figure 5.8a). The beginning of the blend can be just as the right foot is lifting up off of the floor and breaks contact with the floor (Figure 5.8b). The end of the blend can be when the right foot moves ahead of the left foot and is about to come in contact with the floor (Figure 5.8c).

This is basically what you want to blend across. If you decide motion A is first, find the *last* useful step with the left foot on the ground. By useful, we mean you do not want the beginning or the ending of the walk to be incomplete or the mocap subject to turn during the motion. Now look at motion B and find the *first* useful step with the left foot on the ground.

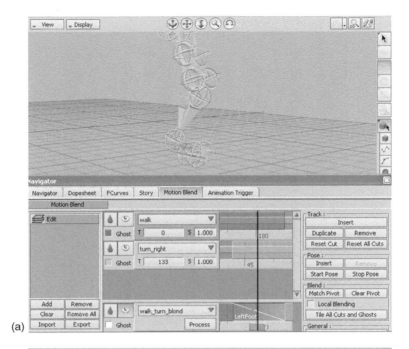

(a)

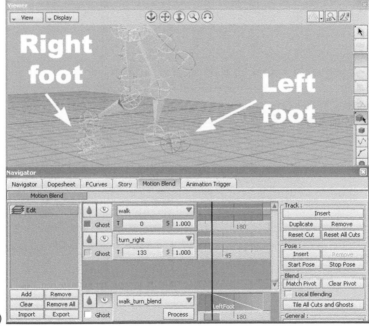

(b)

Figure 5.8 *(a) Right foot in mid-stride passing planted left foot; (b) Beginning of blend with right foot lifting off the floor; and (c) End of blend area with right foot starting to touch the floor again*

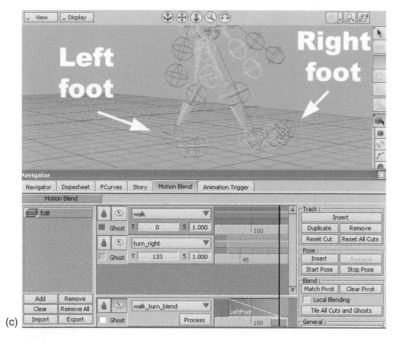

(c)

Figure 5.8 *Continued*

A note of caution: if a character is standing still first and starts walking, the first step is hard to blend into or out of with another motion. From stance to step is a change in body posture and speed. The same can be said for going from walking to standing. That is even more evident with running. It can take several strides to get up to a running speed and several strides to stop, which are good examples of slow in and slow out. During these times, the acceleration of the body is changing drastically as well as the body posture. When you start to run, you throw your upper body forward, and when you are slowing down to stop, you pitch your upper body straight up, even a little back. There are many different postural relations involved in a regular run. For a good run motion, have the capture subject start as far outside the mocap space as possible and have the subject run all the way through without slowing down until the subject is out of the mocap space. This should give a consistent run that will be easy to loop or blend with other motions.

After selecting the segments of motion A and motion B that will be blended, slide the selected segment of motion B under the selected segment of motion A until the selected segments match up. You may have motion before or after your blend that you will not use. That is fine. Focus on the blend section and keep it well defined (Figure 5.9a). Once you match up the selected segments, tighten the blend section even more by specifying the "in" and "out" points. Motion A will end as the left foot is coming down. Motion B will begin as the left foot is coming off the floor. The in and out points let the blending software know where to start blending and where to end (Figure 5.9b).

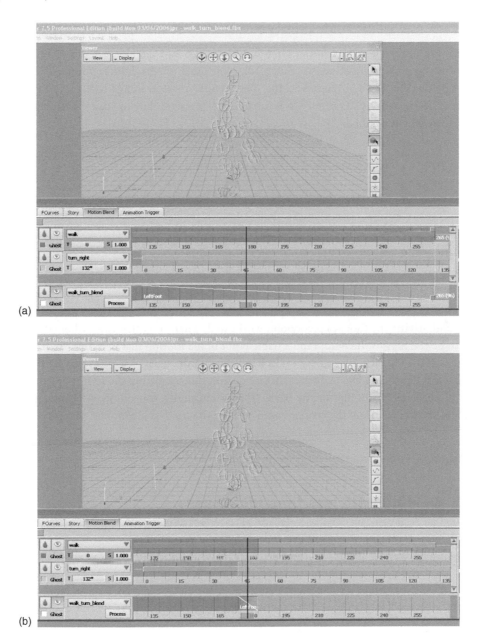

(a)

(b)

Figure 5.9 *(a) Selected sections matched up; and (b) In and end points specified*

5.2.2 Matching positions

It is possible that the two motions are not in the same space. They could be at two different locations. The positions need to be matched; otherwise, a blended skeleton will slide from one location to the other as if it were on roller skates. Use of a "pivot point" or "stable point" allows positions to be matched and a motion to be blended from one to the other in a natural way. Let's look at how positions are matched using a pivot point in MotionBuilder.

In MotionBuilder, select the left foot joint and Alt-drag it into the blended area. The name of the joint shows up in the blended area (Figure 5.10). Now use the "Match Pivot" button (Figure 5.11a). Your skeletons should align (Figure 5.11b). The blended motion should look like one continuous motion. The last thing you need to do is to click the "Process" key to create the resultant motion and the blending is finished.

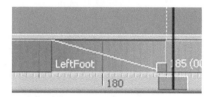

Figure 5.10 *Left foot in blended area*

The foot was selected as the pivot point in the example above. Since the foot is firmly planted on the ground, having the foot as a pivot point works well. There are other cases where the pivot point needs to be the hips or the head. Find a stable point in the motion that is "stable" in relation to the reference. If the character is walking on its hands, one of the hands is a good choice. If the character is in mid-air, choose the body part that is rotating and translating the least, but still going with the motion, possibly the hips or the back. In MotionBuilder you can keep changing your pivot point and see the results of the changes until you get the result that you like.

You can blend any number of motions by continually blending one motion with another. For each added motion, choose a new pivot point to match positions.

5.2.3 Dealing with less than ideal cases

What do you do if you cannot find good places to use as a blending point in the motions that you must blend? For instance, if the backs are in two very different poses throughout two motions, there is no good blending point for them. If the motions were blended anyway, the back would change from one pose to another very unnaturally in the blended motion. Use the hips or one of the back segments as a pivot point. That will give you a much better result than not using a pivot point, although the positions of the feet will be changed wildly and you will need to lock the feet's positions down.

How about speed changes? If you cannot reshoot, use a longer than usual segment for blending. Play around with the length of the blend and fix the feet. Check if the blended motion looks right in

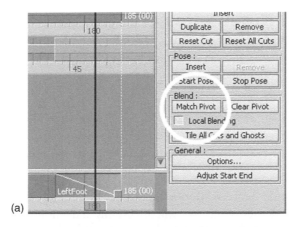

(a)

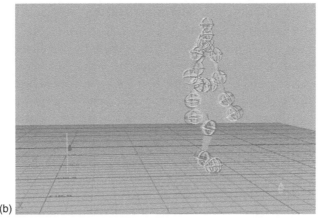

(b)

Figure 5.11 *(a) Match Pivot; and (b) Skeletons are aligned*

terms of a balance. If a person is moving forward and stops, the stance changes, especially in the upper body. If the balance looks wrong, make changes to other parts of the body so that the balance makes sense for the action.

So what if you want to blend a motion of a large person and a motion of a smaller person or vice versa? It is like Dr. Jekyll transforming into a much larger Mr. Hyde. It can be done, but you will have to decide what part of their bodies you want to be stable. Do you want both feet to remain in the same place as the transformation takes place? Do you want the transformation to happen while he is running? Does it happen in the middle of a jump where you can use the hips to be a pivot point and let everything else grow around it? Once you answer these questions, then you will know how to work on the blend.

5.3 Inverse Kinematics

Forward kinematics (FK) is a method of animating a skeleton where the animator specifies and key-frames the position of every joint in the skeleton. In essence, skeletal data is FK. The only difference is that skeletal data has rotational data of all the joints for every frame while FK has rotational data for key-frames only.

Inverse kinematics (IK) is a method of animating a skeleton where the animator specifies only the positions of the end effectors. Software calculates all the rotation angles of the middle joints in the joint chain to reach the position of an end effector. In key-frame animation IK is usually used for animating limbs.

3D applications, such as Maya and MotionBuilder, offer you a tool to edit skeletal data using IK and blending IK and FK animations. We will discuss this in depth in Chapter 8.

IK solvers in MotionBuilder work all at once. MotionBuilder does not currently allow IK passes to be layered (hopefully a future release of the software will). What this means is that it tries to reach all the IK end effectors at once pushing and pulling multiple IK chains around. Because of this, you may fix one problem in one area and create another problem in another area. MotionBuilder does give you the ability to turn on or off parts of the body to be affected by IK but it still tries for an overall fix. If you find one area fighting against another, edit one area of the body, such as the legs, save out the data, and then re-import it to fix another area, such as the arms. A better solution seen in other applications offers layering of IK passes that enables you to solve for the arms, legs, and back independently and all of them at the same time.

In order to use the IK solver in MotionBuilder, you need to characterize a skeleton. Then "plot" the character to the "Control Rig." This creates a control rig on the skeleton that can be manipulated using IK constraints. Next choose the FK/IK option. To get the data saved back to the skeleton, "plot" the motion back onto the skeleton in the Character Settings tab. If your motion does not look exactly the same as it did before plotting, go to the Animation tab at the top of the interface and choose "Plot All," then plot to the skeleton.

5.4 Floor Contact

The floor contact is a useful tool. Various versions of it can be used in different situations, such as keeping a foot on the floor, keeping a hand steadily on an object, and keeping a head constantly looking toward something.

Making convincing contact with the floor or other objects is one of the most important aspects of motion editing. A firm interface between a character's feet and the floor makes the motion look realistic. What are some ways to do this?

Some methods are broader and quicker but give a less accurate result overall. Others are more time consuming but usually give a better look. It is very important that your marker data is as clean as possible. The better the data coming out of your capture session, the less you will have to fix.

One of the major reasons to fix feet is that they have been pulled a little by retargeting applied to another part of the skeleton; retargeting is rarely a perfect fit. The software does not have the same eye as a human, so it works on what is mathematically correct. There are times that mathematical correctness and what we perceive to be correct are two different things, especially when dealing with optimizing or battling IK pulls.

A quick way to fix the feet's contact with the floor is the "floor plane" option in MotionBuilder. When you characterize your skeleton in a T-pose, a set of virtual markers are set up on the hands and feet. You can tell the program to never let them go under the floor plane. The markers are generally at the heel, ball of the foot, and toe (Figure 5.12). Those push the foot up above the floor plane but may not push it up to the same place it was a few frames before. The reason for this is that the marker positions with regard to the floor are computed using only the translations along the vertical axis, that is, the software just pushes everything up until it is above the floor. A foot may need to be rotated, especially when there are multiple markers under the ground plane. However, the floor plane does not rotate the foot. It simply pushes the foot up.

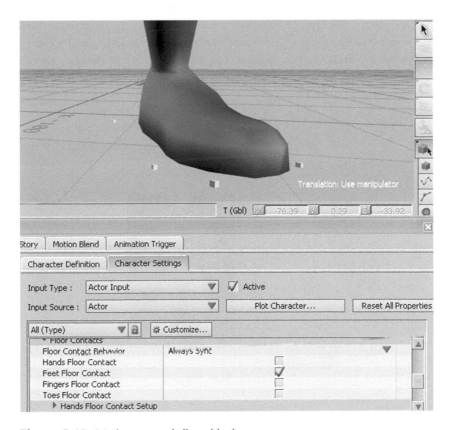

Figure 5.12 *Markers at toe, ball, and heel*

The floor plane option can shorten the slow in and slow out of the foot motion above the floor and make the motion look unnatural if the timing of the step is relatively slow or the foot is moved from far under the floor. Thus, the timing of the foot motion above the floor may need to be edited using IKs.

The floor plane option usually gives you a good starting point but always remember that if the feet are not seen from the camera that will be used in your final rendering, you do not need to clean them up at all.

When in MotionBuilder, make sure to characterize your skeleton in a T-pose. If the character is not characterized in a T-pose, the hand and foot floor plane markers will be skewed and not line up properly.

Now let's talk about a more time-consuming way to clean feet. Let's look at the MotionBuilder's approach with auxiliary effectors. The auxiliary effector is created at a point, let's say at a foot, and then is relative to the global space, so it does not move with the rest of the body (Figure 5.13). You can "target" the foot to lock into this position. To do so, first create an auxiliary effector when the foot is making a firm contact with the floor, or move the foot precisely where you want it to be first and create an auxiliary effector.

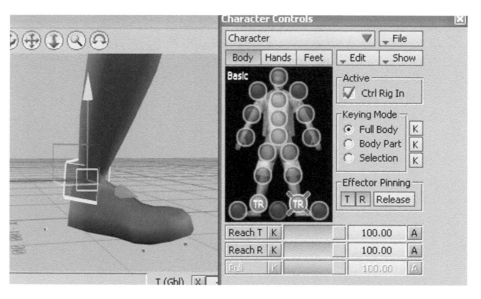

Figure 5.13 *Auxiliary effector created at foot*

When the foot is supposed to be planted, you want the foot to "seek" the auxiliary effector. In other words, you want the foot to be exactly on the effector but not floating around when it should be on the floor. Use slow in and slow out as your character is stepping into and out of the position. Do not let the foot seek the auxiliary effector when it is not close to the effector; otherwise, it will try to pull the entire leg or body toward the point (Figure 5.14).

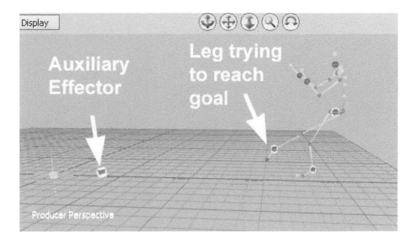

Figure 5.14 *Auxiliary effector pulling entire leg*

So if the foot is working, what about the toe? You can set up an auxiliary effector for the toe or a special constraint to keep the end of the toe from going under the ground. You need to pay attention to the toe when a person is walking, running, etc. and be sure that the toe is in contact with the ground. When the toe and the ball of the foot are the only segments that are touching the ground the auxiliary effector is useful.

There are times when the feet are not the only thing that you need to be concerned with. If a character is leaning against a wall and supporting its weight with its hand you need to have the hand in constant contact with the wall and not floating around. This can also be done with an auxiliary effector.

Another situation is your character running with its hand clutching its side. Usually this type of motion is not dead on and you need to help the hand stay in the correct position. You can create an auxiliary effector so it will move with the right side of the body but not stay in one place. Use the parent–child constraint in order to constrain the auxiliary effector to the spine segment that is closest to where the character's hand should be (Figure 5.15). Look at the movie and .fbx file on the CD.

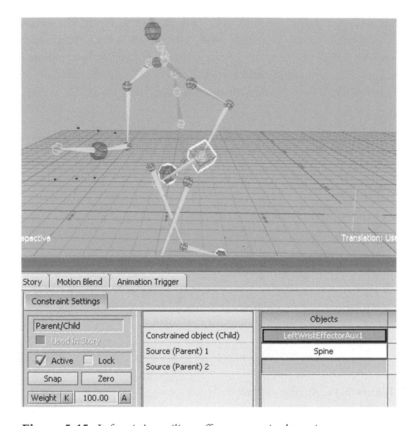

Figure 5.15 *Left wrist's auxiliary effector constrained to spine*

5.5 Rigid Body

A rigid body, such as a sword, a bat, or a football, in relationship to a person presents another unique situation. It can either be attached to the mocapped person and follow the person (Figure 5.16a) or separate and its own entity (Figure 5.16b). When you are working with data that may need to be scaled, be aware that if you attach a rigid body to the skeleton and scale the skeleton, the rigid body will be scaled as well.

If a rigid body is a separate entity, it will make things more difficult when trying to blend motions because you will have to blend the character's motions and then blend the rigid body's motions.

When editing data of a skeleton and a rigid body, first look at the data and determine which has better positional data; the rigid body or the body part that the rigid body is attached to. More often the rigid body has better data but not always. Secondly, have the one with better data move the other.

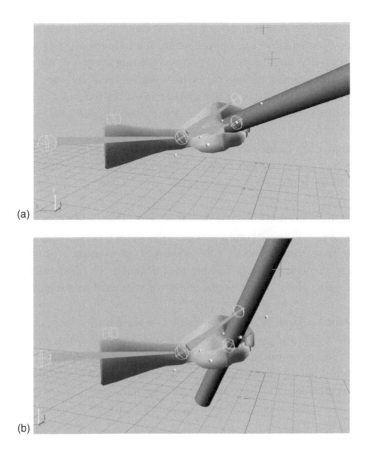

(a)

(b)

Figure 5.16 *(a) Rigid body attached to a person; and (b) Rigid body as a separate entity*

5.6 Looping Motion

One of the most commonly used motions, especially in video games, is the motion loop. In this section we will look at how to loop a motion in MotionBuilder. We will also discuss a few simple preparations that are useful when dealing with video game motions. Even though we are focusing on video games, the same techniques are used for large character simulations that are formulated like a video game and use a number of stock motions.

5.6.1 Getting motion ready

Game engines have certain ways of compiling elements together as specified by the hardware and software. Consult with your programmer and know your application's specifications. The basic rules below should apply to any game engines.

First, create small motions that will be spliced. You do not want excess or wasted motion. An example would be someone drawing a pistol, firing, and holstering the pistol. Break it into three shots. The first shot is drawing the pistol, the second is firing the pistol, and the third is holstering the pistol. The beginnings and endings of the motions should all line up. The shots should flow from one into the other without having any wasted frames. Wasted frames will create a longer wait before the reaction on the screen and therefore slow down the game play.

In the above example, it is normal to never see a character draw a weapon or holster it. Often the weapon just appears. In order to make the action happen as quickly as possible, there are even times when there is no firing animation. Be sure to know what is needed for your project.

Second, align all the motions in the same direction. The general rule is to face the motion down the positive z-axis. Let walking, running, and any other actions move along the positive z-axis and then take all the z-translation out of the motions. If you are working on a walking motion, edit it so it will look as if the character were walking on a treadmill. The forward motion will be added back in when the game player tells the character how far forward to run and for how long.

Thirdly, have standard beginning and ending frames. In the case of a loop, the beginning and ending frames should be the same pose or a pose that is one frame off so that the loop can be repeated seamlessly with no jumps or stutters in the motion. Other motions, such as a transition from a walk to a stop, require a different end pose. A stopped or idle pose is a possibility. Think about what motions need to transition into what motions. An important question is if you really have time to run a transitional motion or if you want to jump directly into the next motion.

5.6.2 Setting up the loop

Let's take a walking motion and turn it into a loop. There are different ways to do this but we are going to step through MotionBuilder to show a general way of creating a loop.

5.6.2.1 Walking down the z-axis

There is a walk motion, walk.fbx file, on the CD. Load it into MotionBuilder. We will use this for the rest of this chapter in setting up a loop. The first thing we need to do is to face the character down the positive z-axis and have the character walk down that axis.

Characterize the skeleton. (See Section 4.2.3 for characterization.) Now plot the character to the Control Rig (Figure 5.17). This will give you a control rig so that the skeleton can be manipulated and the result can be key-framed.

Now we need to swing the entire animation around to make the character walk down the positive z-axis. When the file is opened, the skeleton is walking diagonally across the space because the

Figure 5.17 *Plot to Control Rig*

diagonal distance of a square floor is the longest distance and gives us the most straight line data. We want to turn this entire animation around. We do not want to rotate, translate, and key-frame the first and the last frames of the motion. We will use the Control Rig's Character Ctrl: Reference node that will rotate the character over the entire time frame. It is indicated by a circle located between the character's legs at the bottom of the Character Control panel (Figure 5.18). Once you have this selected, turn on the rotation handle, and rotate the skeleton so that it is pointed down the positive z-axis.

5.6.2.2 Taking out the translation

We are now ready to pull the translation out of the motion. This is called "zeroing out the motion." The first thing we need to do is release the effector pinning on the ankle. If this is not done, the character will try to keep the feet in the original position instead of allowing them to move when the hips translation value is reset to zero. Select each ankle's auxiliary effector in the Character Controls and choose Release under the Effector Pinning section (Figure 5.19).

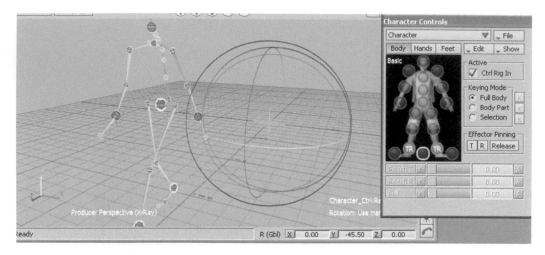

Figure 5.18 *Control Rig Reference*

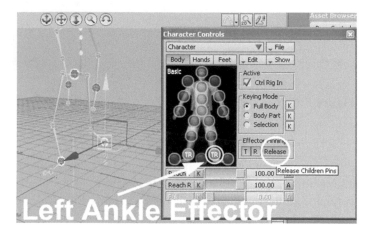

Figure 5.19 *Effector Pinning Release*

Select the translation mode by clicking the translate mode icon on the side or pressing the T key. Go to the first frame of the motion (frame 70 in our case) and select the Hips Effector. Change the z-value in the Global Translate box to 0 (Figure 5.20).

Go to the Key Controls, change from base layer to Layer 1, and press the "Key" button. This will create a zeroed z-translation for frame 70. If you play the motion, it will now start at the center of the space and walk forward. The next step is to do the same thing for the last frame of the motion. Go to frame 200, set the z-translate value to zero, and then key it on Layer 1. Make sure that the Hips Effector is still selected and that you are changing the effector's value. Also make sure that

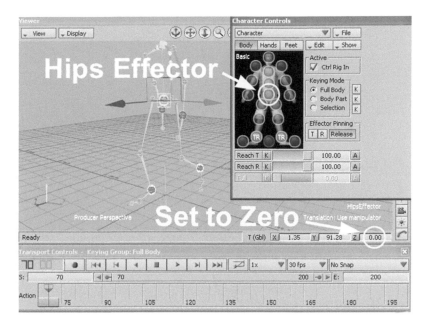

Figure 5.20 *Zeroing the hips*

both key-frames are created on Layer 1. Your character should now look as if it were walking on a moving sidewalk.

If you look at the motion from the side, it is obviously staying around the center of the space and walking in place. If you look at it from the front, however, you see a translation when the character moves to its right. The translation is the most notable in the left footfall around frame 116 and in the right footfall around 131. Not all the steps in the motion are needed for the loop that we are creating. Let's use the frames after the translation, that is, after frame 131.

We need similar frames at the beginning and ending of the loop. Let our selection point be when the right foot is planted and the left foot is moving into a passing position. Let's use frames 150 as the beginning and 180 as the ending. Change the start and end times in the Transport Controls to reflect this (Figure 5.21).

Select the Hip Effector in the Character Controls. Frames 150 and 180 have z-translation values that are no longer zero. Set these to zero and key-frame them as you did above. Play the motion in loop mode. Looking at it from the side, looping looks almost fine, but if you look at it from the front, you can see a twitch between the beginning and ending of the animation. We will look into how you can use poses to take care of this in the next section, but first, let's go to the Animation pull down at the top and select "Plot All (All Properties)." This bakes the key-frames that you created onto the motion and the key-frame indicators disappear. We now have a clean timeline to start using the Pose Controls.

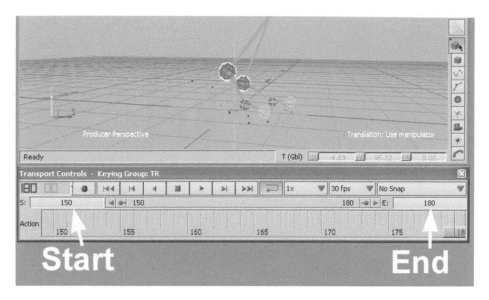

Figure 5.21 *Narrowing the motion*

5.7 Poses

"Pose" is a very useful animation tool in MotionBuilder. It allows you to copy a pose in one frame of a motion and paste it in different frames of the motion. You can then slow in and slow out of the pasted poses. Pose can be used to pull a pose out of mocap data and create consistent ending poses. An example is the idle stance for a video game. This is the generic pose the video game character is in when it is not in motion. Another example is the end pose shared by running motions, walking motions, and jumping motions. If the motions did not have the same ending pose, you would either have to have a large number of transitions, or let motions jump from one to another. You can create end poses in the following way.

5.7.1 Deciding what to use

So what do you use for a pose? That's a good question. Since we have the beginning and ending frames for the loop, let's choose frame 150, the beginning frame. You may decide that frame 180 is better or that you want your walk loop to begin and end with different frames.

5.7.2 Creating a pose

Since we are going to use the pose in frame 150, the timeline indicator should be on frame 150 and frame 150 should be the current frame. Go to the Character Controls and click in the black around the character but not on any specific control. This selects all of the controls except the

Reference node between the feet. Change the Asset Browser to the Pose Controls by selecting the Pose Controls tab (Figure 5.22). (Pose Controls can be found under the Window of the top main menu as well.)

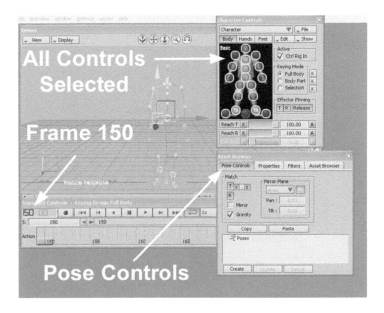

Figure 5.22 *Pose Controls*

In the Pose Controls, click on the "Create" button. A plus sign appears next to the word "Poses." Click on the plus sign. It changes to a minus sign and a Character Pose appears below "Poses." Right click on the Character Pose and change the name to "walk" (Figure 5.23).

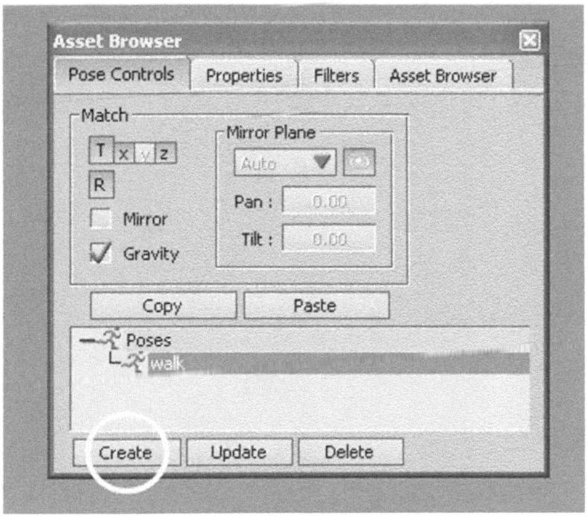

Figure 5.23 *Poses*

We are ready to use the pose at the end of the motion but first make sure that the "Match Translate" option is on (Figure 5.24). If the option is turned off, there will be a translation in the beginning and ending frames that you do not want for a loop. Go to frame 180. Make certain that the pose named "walk" and the controls (except for the Reference node) are selected, and click on the "Paste" button in the Pose Controls. You should be able to see a noticeable change in the skeleton's pose as the pose from frame 150 is inserted into frame 180.

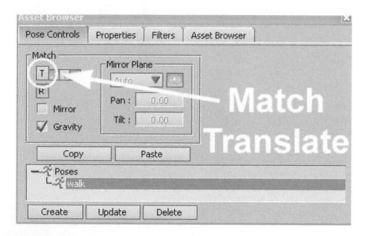

Figure 5.24 *Match Translate*

5.7.3 Key-framing a pose

The pose now needs to be key-framed. Since all of the effectors (except for the Reference nodes) are selected, just add a key by clicking on the Key in the Key Controls (Figure 5.25). After this, play the motion. The loop should look reasonable at this point.

Next we will create a "Zero" key (Figure 5.25). A Zero key forces all the effectors to move back to their original positions. When you key-frame an inserted pose at the end of the motion, it affects

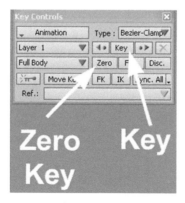

Figure 5.25 *Key Controls*

the entire motion, from the first frame to the last frame. The "Zero" key can localize the effect. Go to frame 170 or 175 (i.e., 5 or 10 frames before the keyed ending frame), and press the "Zero" key. Now the pasted pose affects only the last few frames.

Now all the key-frames are made and the motion is working. Go to the "Animation" tab at the top of MotionBuilder and click "Plot All (All Properties)." This "bakes" all the changes onto the skeleton. Then apply "Plot Character" in the Character Settings so that you can export the skeleton back to your 3D package of choice.

Hopefully you have a good idea about skeletons, retargeting, blending, IK, floor contact, rigid body, looping, and poses. In various data editing software packages these procedures are implemented in different ways. Know what each procedure does and what needs to be done, and try tools to find a way to get the desired results.

In the next few chapters we are going to look at the pipelines for props, characters, hands, and faces.

6 Data Application — Intro Level: Props

We looked at how data is cleaned and edited in the previous two chapters. You are probably eager to know how to apply data to your 3D character. But before showing you how to apply human motion data to a 3D character let's talk about much simpler cases. We will be mostly talking about marker placements on props for optical systems in this chapter but the same principles work with human capture subjects. Therefore, this chapter should help you understand effective ways to place markers on humans as well as props.

6.1 A Stick with Two Markers

Let's consider capturing a stick as a prop. Suppose that you place a marker at each end of the stick. You give the "markered" stick to a performer, ask the performer to move it around, and capture the motion of the stick. Let's disregard the motion of the performer for simplification for now. Process the data as described in Chapter 3 and import it into your 3D application. Go to the first frame where you can see the two markers. Create a joint j_1 at one marker m_1, create another joint j_2 at the other marker m_2, and a bone between the two joints. Let the position of joint j_1 be constrained by the position of marker m_1 and the position of joint j_2 by the position of marker m_2. Play back the animation. You can see the simple rig (a bone with two joints) being animated by the mocap data. If you like, create a skin geometry, for example a cylinder, and bind it to the rig.

6.1.1 When it fails: Occlusion

The approach above fails in many ways. The most obvious cause is occlusion. When one of the markers is occluded by the performer there will be only one marker to be tracked. If you play back the animation you will see that the tracked marker moves one end of the stick (i.e., one joint in the rig) to the correct position for each frame but the other end of the stick does not move while the other marker is occluded. Remember that the optical system gives you a marker's position but not its orientation. Mathematically two points are required to define a straight line in 3D space (Figure 6.1). Having one marker is like having just one point. It is impossible to orient the stick with one marker.

If an occlusion happens for a short period of time, you may be able to fill in the gap by applying a spline interpolation or some other method as we discussed in Chapter 3. But if the occlusion period is long, there isn't much you can do, besides painstakingly filling in the gap by key-framing or giving up on the data and recapturing.

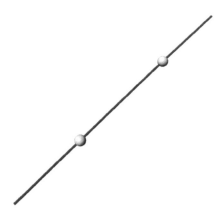

Figure 6.1 *Two points define a line*

When both of the markers are occluded for a period of time and reappear, your application that reconstructs trajectories may swap them in the middle of the sequence by mislabeling them. For instance, the marker that's labeled as m_1 before an occlusion can be labeled as m_2 after the occlusion and the marker that's labeled as m_2 before the occlusion can be labeled as m_1. This will be an issue if both ends of the stick are not identical. Imagine a crook cane. You want your 3D character to hold the cane's crook, but not the tip (Figure 6.2).

Figure 6.2 *Character with a crook cane*

6.1.2 When it fails: Rotation

We've looked at the possibility of two markers being mislabeled and swapped after occlusion. Marker swapping can happen when the stick is quickly rotated as well. For instance, if a stick is swirled around like a baton when your capture rate is set relatively low, then your application will be confused, so will you.

6.2 A Stick with Three Markers

Let's try placing three markers on a prop stick this time. When one marker is occluded, there will be two markers to determine the orientation of the stick. Also by having three markers we can capture the stick's bending if the stick is flexible enough to bend. So, we can expect a better result with three markers than two markers. Now the question is "What is the best way to place three markers?"

6.2.1 Three markers with equal distances

Let's place the first marker m_1 near one end of a stick, the second marker m_2 at the middle of the stick, and the third marker m_3, at the other end of the stick so that the distance between m_1 and m_2 and the distance between m_2 and m_3 are equal (see Figure 6.3).

Figure 6.3 *Equal distances between markers*

Capture motion data while a performer is moving the stick around. Process the data and import it into your 3D application. Go to the first frame where you can see the three markers. Create joint j_1 at marker m_1, joint j_2 at marker m_2, joint j_3 at marker m_3, a bone between j_1 and j_2, and another between j_2 and j_3. Let the position of joint j_1 be constrained by the position of marker m_1, the position of joint j_2 by the position of marker m_2, and the position of joint j_3 by the position of marker m_3. Play back the animation. You can see the rig with three joints being animated by the mocap data. If you like, create a skin geometry and bind it to the rig.

With three markers you are less likely to lose the orientation of the stick due to occlusion than with two markers. However, when the stick is rotated like a baton, either you or your application has no way to know which end marker was at which end of the stick. The confusion can be avoided by placing the middle marker m_2 at an off-set position so that the distance between m_1 and m_2 and the distance between m_2 and m_3 are not equal (see Figure 6.4). By having unequal distances between the markers you and your application can immediately identify which marker is which.

6.2.2 Three markers on a single straight line

Now let's think about a cane and the orientation of its crook. Suppose that we want the crook to be oriented correctly so that it stays in a 3D character's hand, but the marker set in Figure 6.4 does not give you the orientation of the crook. You will encounter a similar problem if a prop is a rifle. The rifle's muzzle needs to point in the direction that it's aiming. At the same time, the rifle's butt needs to be oriented correctly so that the projection side is up and the trigger is down. However, again a marker set like the one in Figure 6.4 will not give you the data that orients the rifle's butt properly. Let's understand why.

Figure 6.4 *Better position for the middle marker*

In the previous section, we mentioned that two points can define a line in 3D space (Figure 6.1). Three "non-linear" points in 3D space can define two lines and a plane which a cane or a rifle can be laid on (Figure 6.5). Hence, a plane defined by three non-linear points can orient a cane's handle and a rifle's butt as well as the rest of them. However, three "linear" points can define only a single straight line and no plane (Figure 6.6).

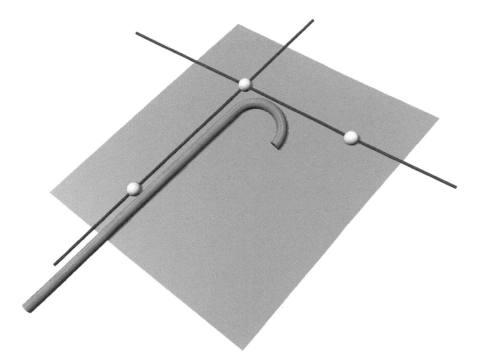

Figure 6.5 *Three non-linear points define two lines and a plane*

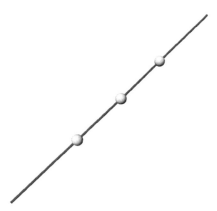

Figure 6.6 *Three linear points define a line*

We have three markers but they are on a straight line. When the markers are placed on a prop in such a way that they all line up on a single straight line, the marker data lets us reconstruct the straight line that the markers were on, but not a plane. Thus while the cane's tip and the rifle's muzzle can point at the right direction, neither the cane's crook nor the rifle's butt can be oriented properly with three markers on a straight line.

6.2.3 Placement of three markers that works

In the previous two sections, we learned that neither three markers with equal distances nor three markers on a straight line is an effective way to place a trio of markers on a prop. The best way to place three markers is to create unequal distances among them and not to let them line up on a straight line (see Figure 6.7).

Figure 6.7 *Best way to place three markers*

Although the marker set in Figure 6.7 is better than the one in Figure 6.3 or Figure 6.4, it is still not ideal. If one marker is occluded the orientation of the prop will be partially lost. If two markers are occluded, the orientation will be completely lost. In practice use at least four markers. With six or seven markers you can expect better results. If a prop is flexible, you need more markers. (We will talk about marker setup for flexible objects in the next section.)

If you are a Maya user, try the following experiment in Maya to understand how three markers can orient an object. Create a polygon and three joints, j_1, j_2, and j_3. (Three joints can be in a hierarchy or disjointed. Either way works. Also they can be locators or anything else that you can easily select and move.) Point constrain joint j_1 to the polygon (i.e., let the position of the polygon be constrained by the position of joint j_1). Aim constrain joint j_2 to the polygon with "Object Up" option as "World Up Type" and joint j_3 as "World Up Object." Move the three joints around individually to see how the polygon's orientation changes. Display the polygon's local rotation axes. You can see that the origin of the polygon's local space keeps moving with joint j_1; the x-axis of the polygon's local space keeps trying to point at joint j_2; and the y-axis of the polygon's local space keeps trying to point at joint j_3. If you point constrain a marker to each joint you can have the polygon animated by mocap marker data. And this method works not just with polygons but also with 3D objects.

We have looked at simple placements of two or three markers on rigid objects in this section. The same principle works for marker sets for human and animal capture subjects. For instance, if we want to capture the position of a wrist (but do not care about its orientation), one marker on the wrist is sufficient (Figure 6.8a). However, if we want to capture the wrist's twisting as well as its position, we must place two markers on the wrist in such way that the wrist markers and an elbow marker make a triangle, not a single straight line (Figure 6.8b).

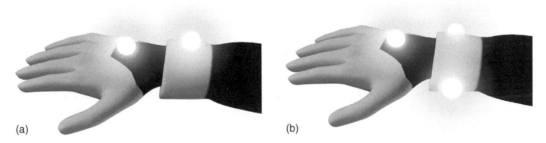

(a) (b)

Figure 6.8 *(a) One wrist marker; and (b) Two wrist markers*

6.3 Flexible Objects

If you have a flexible rod, such as a golf club, or better yet, something more like a pool noodle, you will need multiple markers to define how the object bends. So, how do you figure out how many markers are needed to capture a flexible object? The answer is: it's hard to tell until you capture it a few times. The best way is to give it a few tries but we'll try to give you a few tips on how to place markers on flexible objects.

First of all, you need to know how flexible the object is. You want to be able to capture the dynamic motion of the object, but you don't want to apply too many markers or have them too close to each other. When we say flexible, we mean something that won't coil up or is extremely flexible such as a rope, fishing line, or whip, although you can track all of these.

Bend the object as much as you can and try to imagine how many straight line segments you can put into the object that fit within the object's curvature. Below is both an example of a rather stiff object that only slightly bends (Figure 6.9a) and a more flexible object that bends more (Figure 6.9b). The stiffer object may only require three segments, whereas the more flexible object may require seven segments.

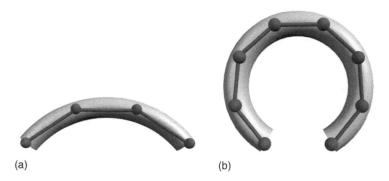

(a)   (b)

Figure 6.9 *(a) More rigid object requires fewer segments; and (b) More flexible object requires more segments*

After making an educated guess about how many segments to approximate the object's deformation, you need to think about the number of markers. At least three markers are needed for a rigid object as we discussed in the previous section. You need more for a flexible object. You want to place at least two markers in each segment of a flexible object. For the less flexible object with three segments in Figure 6.9a, you may place three markers (including the base marker, which we will talk about shortly) on the first segment and two on each of the other two segments for a total of seven markers. For the more flexible object with seven segments in Figure 6.9b, you may place three on the first segment and two on each of the other six segments for a total of fifteen markers.

Stagger markers along the length of the object. Staggering prevents markers from lining up on a single straight line and also makes it easier to identify the markers with a glance.

The next thing you need to think about is if the object comes in contact with people and if it does where the contact point will be. Before we get into this further let us define the "base marker" as the marker on an object that is selected to provide data for the overall translation of the object. If the object never comes in contact with anyone, you can place the base marker at any position along the object. If you know one end of the object will be picked up and spun around, place the base marker on that end. If both ends will be picked, you have to decide on which end the base marker should be placed. Having just one marker providing the object's translation becomes very important when you need to edit the object's motion, specifically in reference to the motions of other objects in the scene. Keeping the base marker at the same location of the object throughout the shot allows you to see the object move in relationship to other objects (e.g., the hand that is holding the object) and gives you the ability to translate the object with as much control as possible. Above is true for rigid objects as well.

Moreover, if the object will be picked up by someone make sure that the markers are placed away from where the hand will grab the object. Even if the person interacting with the object has no markers on her/him, her/his hand can still cover up some markers.

Now let's talk about how you can apply motion data of a flexible prop to a 3D model of the prop. If your mocap software can give you skeletal data, create a skeleton for a flexible prop shaped like a rod with a single chain of joints. The number of the bones in the skeleton should be same as the number of segments that you came up with for the prop. The root joint of the skeleton should be placed where the prop's base marker is. To bring motion into your 3D application, import skeletal data into the file that contains the skeleton with skin bound to it.

If your mocap software does not give you skeletal or rotational data, bring marker data into your 3D application and create a spline curve using the marker positions as the positions of the control vertices (or end points) of the curve. The first control vertex (i.e., the beginning) of the spline curve should be where the base marker is. The IK Spline tool in Maya is a nice tool to apply positional data of a long flexible object to its 3D model. The tool can deform the skin bound to the skeleton by changing the shape of the joint chain using a spline curve. Applying an IK spline to your skeleton using the spline curve controlled by the marker data will deform and animate a 3D model of the prop.

If your flexible prop is not shaped like a rod but more like a pillow, you need to segment it into a grid. Place a marker at each vertex of the grid, including the internal vertices. The first two facial rigging methods in Chapter 10 work well with positional data captured using markers in a grid configuration.

Let's look at more challenging data applications in the next two chapters.

7 Data Application — Intermediate Level: Decomposing and Composing Motions

This chapter is about tearing motions apart and reusing them. For example, suppose that a game character has 50 different walking motions and 30 of the 50 motions have identical lower body motions but different upper body motions. To minimize the data size and the memory that are required to play the game, the 30 motions can be divided into the upper body motions and the lower body motions. Having one lower body motion (i.e., walking) shared by the 30 different upper body motions (e.g., reloading a rifle, pulling a trigger, and throwing a grenade), only one lower body motion is necessary for the 30 walking motions.

7.1 Mapping Multiple Motions

If decomposed/composed motions need to be prepared for a game on a game engine, programming knowledge specific to the engine is required. We will show you generic ways to decompose and compose motions using the example above and Maya's constraints and Trax editor.

7.1.1 Decomposing and composing upper and lower body motions

The first step for creating a lower body motion without an upper body motion is bringing in a normal walking motion with no extraneous upper body motion. The easiest way to create the lower body walking motion, without the upper body motion, is to select and delete all the joints above the hip joint in the character's skeleton (Figure 7.1a and b). Just a pair of legs should be walking around. Save this out into a file with a file name, such as "walking_lower.mb."

The second step is to create an upper body motion without a lower body motion. Import one of the full-body motions with the upper body motions that you want to keep. Delete all the joints in the legs. The hip joint and everything above the hip joint should be left (Figure 7.2a and b).

The third step is to remove the translation and rotation from the upper body's hip joint since the translation and rotation of the lower body's hip joint will be used. Select the x-, y-, and z-translations and rotations of the upper body's hip joint in the channel box (Figure 7.3), right click on Channels, and then choose "Break Connections". The color of the boxes containing the translation and rotation values should change from orange to white. Play the motion. The lower body is walking and the upper is moving above the hips but not going anywhere.

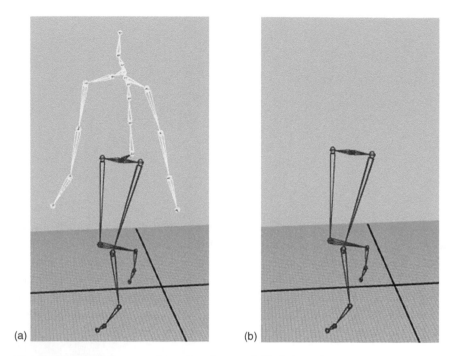

Figure 7.1 *(a) Select everything above hips; and (b) Delete selection*

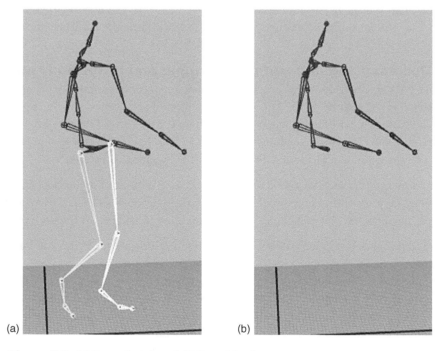

Figure 7.2 *(a) Legs selected; and (b) Legs deleted*

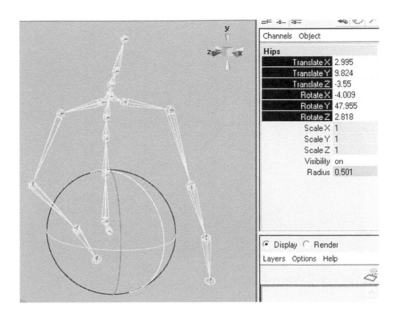

Figure 7.3 *Breaking connections*

The fourth step is to let the lower body move the upper body using point constraint. Select the lower body's hip joint and shift-select the upper body's hip joint. Apply point constraint with the default settings. The position of the upper body's hip joint is now constrained by the position of the lower body's hip joint. The upper body should be moving with the lower body. However, the upper body is facing the wrong direction (Figure 7.4a).

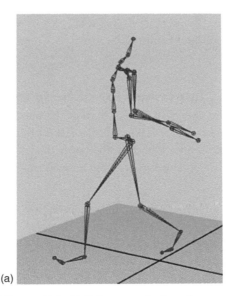

(a)

Figure 7.4 *(a) Point constrained; and (b) Point and orient constrained*

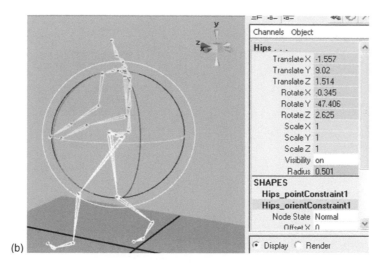

(b)

Figure 7.4 *Continued*

The fifth step is to let the lower body orient the upper body using orient constraint. Select the lower body's hip joint and shift-select the upper body's hip joint. Apply orient constraint with the default settings. The orientation of the upper body's hip joint is now constrained by the orientation of the lower body's hip joint. The upper body should be facing the direction that the lower body is facing (Figure 7.4b).

The last two steps can be done using other Maya's tools, such as the Expression editor, Script editor, or Connection editor, instead of constraints.

7.1.2 Synchronizing upper and lower body motions

Let's keep using the example above. In the last section the upper body motion and lower body motion are decomposed and composed. The positions and orientations of the upper and lower bodies are matched using constraints. However, the upper and lower bodies may not be moving in a synchronized manner yet. For instance, the timings in the counter movement of the arms and legs may be off.

To synchronize the upper and lower body motions, select all the keys for the upper body motion and carefully move them in time until the arm movement and leg movement are aligned. This can be done using Maya's Trax editor as well. Since the Trax editor works with character sets (not with keys), character sets for the upper and lower bodies must be created. Create a character set for the upper body by selecting all the joints in the upper body's skeleton and another set for the lower body by selecting all the joints in the lower body's skeleton (Figure 7.5).

Create animation clips and bring them into Trax (Figure 7.6a). The clips appear as tracks that can be moved in time (Figure 7.6b). Move the clips in time until the upper body motion and lower body motion are synchronized. (Read the Maya manual for more detailed usage of the Trax editor.)

One useful process that you can take before exporting motions out of MotionBuilder, or any other motion editing tool, is to align all motions so that a character is in identical poses at a selected frame number. For instance, move motions in time so that at frame 30 in all the motions the left foot of the character is planted and the right foot is halfway through being lifted. It is a time-consuming

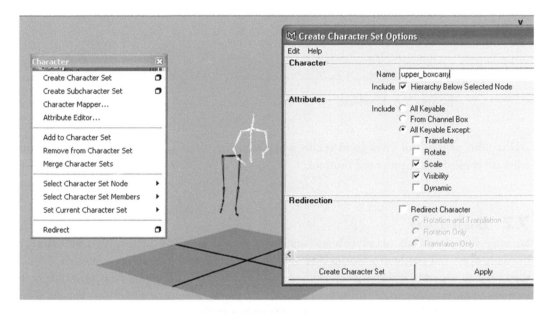

Figure 7.5 *Selecting joints and creating a character set*

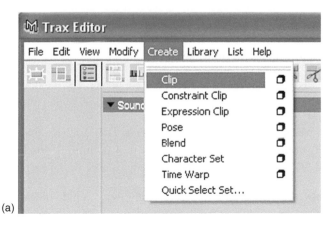

Figure 7.6 *(a) Creating a clip in Trax; and (b) Tracks in Trax*

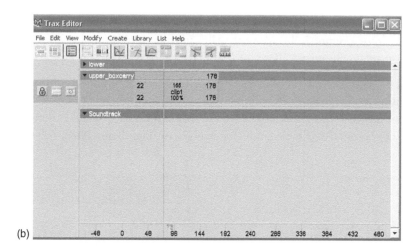

Figure 7.6 *Continued*

and tedious process but gives good results when blending, decomposing/composing, and other edits are applied to the processed motions.

7.2 Balance

So why should we talk about balance, and what do we mean by balance in relationship to breaking apart and combining mocap data? Unless your characters are floating weightless in space, they need to appear as if they have normal everyday forces applied to them. Gravity is usually the greatest and most recognizable influence. The force of gravity is evident when you capture. When you start mixing and matching data the center of gravity can change. The computer programs don't take the effects of gravity into effect when you want to mix motions together.

A character has balance when its weight is equally distributed between its feet. When it goes outside of its center of gravity, the character, without adjustments, would fall over. There are certain exceptions to this, for example, leaning against an object.

If you balance on one leg, your means to stabilize is cut in half. Try standing on one foot and extending your arms in the same or different directions. See what makes you go off balance.

For good demonstrations of balance, watch dancers. Dancers have excellent poise and balance. Their great bodily control can be seen in the way they position themselves and carry themselves. Cirque Du Soleil DVDs are great references to study balance in the extreme.

If the balance looks off it is wrong. There is not really a way to hide or fake it. Make sure that when you are adding or merging different motions together you get the proper balance. The body in the composed motion should appear counterbalancing itself to maintain its equilibrium.

If you are working on a 4-armed creature, it is important to make sure that the upper and lower arms balance each other out. If all four arms are out in front of the character, the back needs to be arched back to keep it in balance. This avoids a mechanical posture.

Another thought about the creature in the above example is to increase its base of stability. Widening the stance, increasing the size of the feet, or adding more legs can increase the base of stability. When you are creating imaginative 3D characters use references just as you do when you are creating realistic 3D characters. Study the forms and movements of elephants' legs if you are designing on a heavy animal. Studying the anatomy and morphology of animals helps you design the look and movement of imaginary creatures that will suspend disbelief. (Morphology is the study of forms and functions.) A trip to the local zoo with a sketchbook and a video camera is time well spent. Call the zoo ahead of time to see when the animals are most active.

7.3 Breaking Motion Apart

How often do you see mocap data pulled apart and used in pieces for different purposes? Many people overlook these possibilities because mocap systems are primarily geared toward full-body motion capture. Only a few look at what their system can do beyond its primary purpose.

What if you want to have an arm sticking out of a wall and motioning to someone? Something to think about, before your capture, is how the arm movement will affect the rest of the body. When a person waves someone over, the entire body moves. Ask the actor to try to isolate the arm. If it does not work, when you pull the arm off, include the shoulder and part of the back. It would look strange if you pulled the arm movement alone because the arm is part of a hierarchy, so it depends on the joints above the arm joints in the hierarchy for some of its movements.

Large motion in the higher parts of the hierarchy often drastically affect where the body parts lower in the hierarchy will end up. Try to reach straight up in the air without moving your shoulder or rotating your back. Then use your shoulder and rotate your back so you can reach as high as possible. You will see the back and shoulder drastically affect how far the arm can reach and what orientation it's in.

Give in-depth consideration to as many of these questions as possible in preproduction. Provide your capture subject with detailed acting direction. It will make your life easier down the road.

7.3.1 When you don't need all the motion

There are times you have too much motion. This usually happens when your motions were not thoroughly thought out before you captured them. Let's use the arm sticking out of the wall as an example. Your performer was very active and threw the whole body into the performance. This is usually what you want but, in this case, you wanted less and the shoot is over. It would be too costly to reshoot just one thing. You now need to create what you need from the data you have.

The first thing to do is decide what part of the motion is really important. The arm consists of the upper arm, lower arm, and hand. The hand makes the motion that really expresses the intent of the motion. So the hand position and movement should remain intact and be used to drive the rest of the arm. The lower arm is important too because you'd like to preserve the general direction of the elbow if at all possible. The upper arm motion can be changed the most.

If we take all the arm motion off and just stick the upper arm in the wall with no spine or shoulder motion, it looks jerky. What we would really like to do is preserve the hand and lower arm motion and mute the upper arm. Let's look at one way to do this in Maya.

What we will do in Maya is remove all the motion from the upper arm and the lower arm and replace the deleted motion with an inverse kinematics (IK) chain that will follow the original motion of the hand, pulling the upper arm and lower arm with it. Where the IK chain bends will be controlled by the position of the elbow; the goal is to make it look good enough to get the point across.

The first step is to track the position of the hand. Run the script named "armpos" on the CD. The script does three things. First: it stores the position and rotation of the hand at every frame. Second: it applies the data to a cube. Third: it creates a locator a short distance away from the elbow. The locator, "elbow locator," moves with the elbow.

The second step is to go to the first frame of the motion and delete the motion of the lower arm, the upper arm, and the hand. You can do so by breaking all the connections of their translation and rotation attributes.

The third step is to create an IK from the upper arm to the hand (Figure 7.7a) using the rotate plane solver. Set a preferred angle for the joint before creating the IK handle. Select the elbow locator, shift-select the IK handle, and apply pole vector constraint. This forces the IK chain to bend into the direction of the elbow locator (Figure 7.7b). Then parent the IK handle to the cube that has the hand data (Figure 7.7c).

The fourth step is to orient the hand geometry using the hand data attached to the cube. Select the cube, shift-select the object "RightHand," and apply orient constraint.

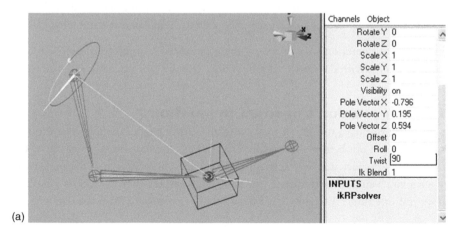

Figure 7.7 *(a) IK handle between upper arm and hand; (b) Elbow locator as pole vector; and (c) IK handle parented under "hand"*

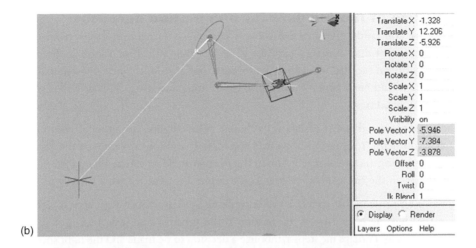

Translate X	-1.328
Translate Y	12.206
Translate Z	-5.926
Rotate X	0
Rotate Y	0
Rotate Z	0
Scale X	1
Scale Y	1
Scale Z	1
Visibility	on
Pole Vector X	-5.946
Pole Vector Y	-7.384
Pole Vector Z	-3.878
Offset	0
Roll	0
Twist	0
Ik Blend	1

(b)

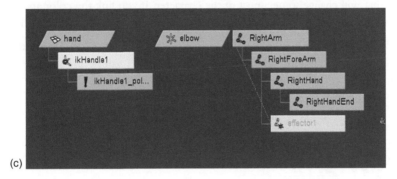

(c)

Figure 7.7 *Continued*

The fifth step is to make the upper arm and lower arm long enough to reach the IK handle by moving RightForeArm and RightHand joints.

Because the arm was changed in length to get a proper IK solution, you may need to scale the motion down. The reason why you have to change the arm length is that we deleted the back and shoulder motion that would have helped the hand reach its goal. Changing the arm length is one way to make up for this. If you already have skin geometry made for the arm, then you probably do not want to remodel it. Try to do this without lengthening the arm. Just move your entire arm closer to where the hand motion happens. You may need to try it a few times to get it to work, but this can be an acceptable solution that requires no adjustment on skin geometry.

The last step is if you want turn the IK data into forward kinematics (FK) data, bake the transformations onto the arm joints. Select the three right arm joints and select Keys under Edit. Then select Bake Simulation. This will set all the rotations, translations, and other attributes as keyframes and the arm motion will no longer be dependent on the IK handle.

7.3.2 Re-use of motion data for non-motion purposes

Our bodies are constantly giving us feedback. We are always getting stimulus from our environment, and because of that, it is very hard for us to be 100% still. If you look at the graphs of the skeletal data, you can see that all of the rotations are not as smooth and controlled as key-framed animation data would be. There is noise and jitter in human motion. This provides you with a lot of interesting possibilities.

One place you can use this sort of data is to change colors, lights, and other attributes and entities in a 3D environment. You may need to find the highest and lowest values of the piece of mocap data that you're using and normalize it. Then the normalized data can be piped into the color value of a shader or even the transparency value. You can use it to create a wide variety of changes.

Let's say you have some mocap data of a person walking up steps. You want a light to get brighter the farther up the steps he goes. Maybe the motion is the person being indecisive, going up and then down, then up again. Perhaps the steps symbolize a decision to be made and the light shows how close he is to a decision that will change some aspect of the character. If you link the color and intensity of the light to the y-translation value of the character's hips, now mocap data is controlling the light.

You can also apply joint rotations to RGB or other shader values (Figure 7.8). To do this, select a joint and map the joint's x-, y-, and z-rotation values to the RGB values of a shader. The shader changes the color as the joint rotates (see also attached Maya file).

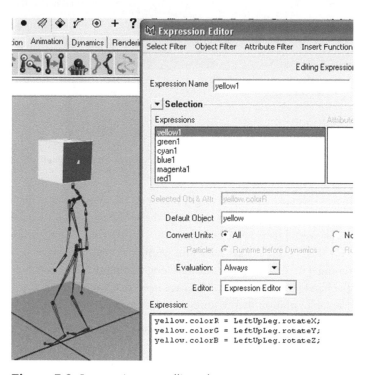

Figure 7.8 *Leg rotations controlling colors*

One last example is to try to use the motion to control small little movements. Take some of the mocap data, especially the jittery parts of rotations, and apply it to models of bugs. Use the data to make their antennae twitch or have extra sets of arms or feelers move erratically.

There is no end to creative uses for mocap data. Always let your creativity flow and try not to get too caught up in all the technical button pushing. Motion capture is a tool for artists to create with and it is not something that takes the place of artists.

8 Data Application — Advanced Level: Integrating Data with Character Rigs

Motion capture is a good tool all by itself but being able to integrate traditional rigging and key-framing on top of it makes it even better. Learning how to integrate motion capture data onto a fully rigged character is an important part of motion capture. Let us look at mixing motion capture into rigging and key-frame animation.

8.1 Mocap as Forward Kinematics Animation

In motion capture we generally expect a direct representation of the capture subject expressed in a digitally created character. Whatever the mocap performer does, you want to see its digital counterpart do the exact same thing. The marker data is transferred to the skeleton in order to replicate the performance.

You may be asking, "What is forward kinematics (FK)?" With FK you position all the joints in a skeleton in one frame, set keys on all the joints, move to another frame, position all the joints again, and key-frame again until you have the motion that you want (Figure 8.1).

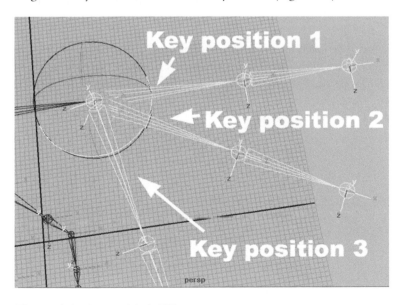

Figure 8.1 *An arm joint's FK*

FK key-framing on every frame generates mocap data. Thus, in each frame the marker point data moves the skeleton into a position and then the joint rotation angles are key-framed. For every position in time, there are a set of directly corresponding rotations and translations. The main difference between mocap and regular FK key-framing is that in mocap every single frame of the motion is a key-frame, whereas for FK a key-frame is created wherever it's needed. The fact that every frame is a key-frame for mocap makes mocap data incredibly large. This is why editing motion capture data can be difficult.

There are times when you can use motion capture data for FK, inverse kinematics (IK), or both FK and IK. Sometimes not using mocap data at all is the best solution. Never get locked into only one way of solving a problem, expressing your creativity, or visually telling a story.

To apply FK data to our skeleton, let's look at MotionBuilder, a stand-alone package that's made for this purpose. MotionBuilder applies marker data (translational data) to a pre-made skeleton by generating rotational and translational data for the joints in the skeleton. It also offers quite a bit of flexibility. A minimum number of back and head markers can drive a fairly large number of back and neck segments (currently 10 of each).

The skeleton needs to be a continuous hierarchy skeleton, not a broken hierarchy or a collection of discrete pieces of hierarchies. We want to start with a totally clean skeleton without any rigging, key-framing, or constraints for this process. We will integrate the IK and FK skeletons later on. Export a basic skeleton out of Maya as an .fbx file.

In MotionBuilder, import marker data and attach it to an Actor. Then import your skeleton into the scene and "characterize" your skeleton. Once this is done use the Actor's motion as the input

There are times when a character motion, after plotting, looks different from before plotting. Save everything into a file before you plot a character. Go under the Animation tab and choose "Plot All (All Properties)" (Figure 8.2), then use "Plot Character" to get all the motion onto the skeleton.

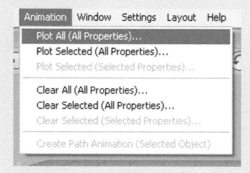

Figure 8.2 *Plot All option*

for your Character's motion. Then plot the Character to get all the motion onto your skeleton. (This is all covered in Chapter 4, review if needed.)

Once this is finished, select the skeleton and export it into an .fbx file (animation only). If you use "Save As," it will save your animation, marker data, and anything else in the scene into an .fbx file. That can be helpful at times but for our current pipeline, let's export animation data only into an .fbx file.

In Maya open the scene file that you exported the skeleton out of. Then import the .fbx file you just exported out of MotionBuilder. An option dialog box will pop up. Use the "Merge" option (Figure 8.3). (There may be times when you use "Convert incoming deforming Nulls to Joints" when there are nodes in the hierarchy that are not used.) You should now see your character moving in Maya.

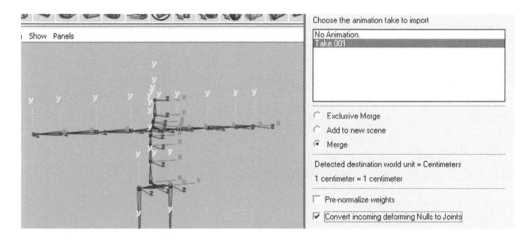

Figure 8.3 *Merging an .fbx file into Maya*

All of these data on the skeleton are FK data but there are a lot of them. MotionBuilder gives you several options on how much data to export. 30 Hz data (i.e., 30 frames per second (fps)) is a standard frame rate, but your end application may want to use another frame rate. Try to match the frame rate of your final render, if at all possible, to the destination fps rate.

Because there is so much motion data, we use applications that allow us to edit and apply the motion in layers, such as MotionBuilder. The density of data makes it imperative that we are able to layer changes on top of the data in order to manipulate it. This is another reason to have a mocap skeleton, an IK skeleton, and another clean FK skeleton. Those make creating your final animation easier.

8.2 Key-frame Animation with Inverse Kinematics

Here we are going to look at the basics of IK. This is not meant to be a rigging book, but you need some understanding of how to set up IK, and it is fairly usual that you want to have an FK/IK rig setup in whatever application you are using.

8.2.1 Key-framing

In 3D animation, key-framing is the process of keying attributes of entities in key-frames (Figure 8.4). Attributes can be translation, rotation, scaling, colors, and any other variables. Entities are often geometries or joints but can be lights, cameras, shaders, and anything else in a scene. If you are an animator and already familiar with key-framing, you can skip this intro to key-framing.

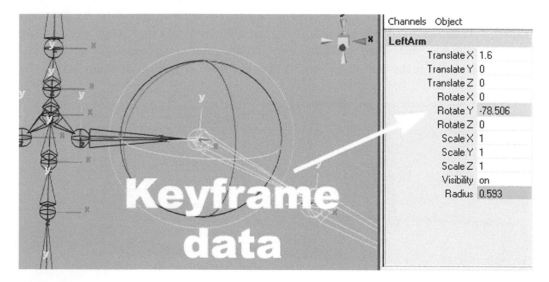

Figure 8.4 *Key-framing the* y-*rotation of LeftArm*

The usual way to set key-frames is to place an object at the primary (or "key") motion points on a path of action. The bouncing ball is probably the most common example that is used to demonstrate fundamental key-framing techniques. To create a bouncing ball animation, have a ball fall in the air, hit the ground, bounce back into the air, and hit the ground again (Figure 8.5). While the ball is translating through the space, the height of the bounce needs to decrease after each bounce in order to mimic the natural effect of gravity. The ball also rotates as it bounces. If the principle of squash and stretch is applied, the ball also needs to have deformations. All of these key moments for translation, rotation, and deformation are key-framed.

Once you have key-frames, the animation software will fill in the frames between these keys with "inbetweens." Let's say you start with a ball on the ground at frame 1. You key the translation. At frame 20 you position the ball in the air and then key the translation there. This will cause the ball to move directly from frame 1 to 20. Even though you didn't key any data in frame 10, you'll see the ball one half of the way between the positions in frames 1 and 20. Another thing you can do is change the arc of the curve that interpolates data between frames 1 and 20. This can give a little more hesitation at the beginning or end and make the graph of your translation not look so linear.

You start by roughing in the translations and then add the rotations and the scaling to give the overall effect of the ball realistically bouncing. If you are interested in more about animation, there are

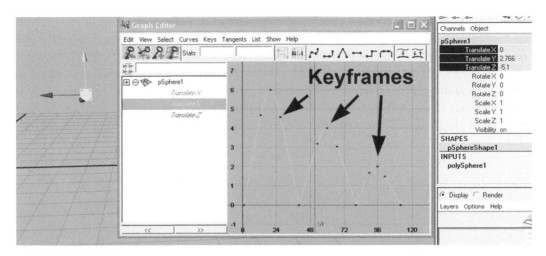

Figure 8.5 *Key-frames on a graph*

several great web sites and books. Anything more than this starts to go beyond the realm of this book. We only want to demonstrate what a key-frame is and how it is used. In mocap data, there are no inbetweens because, as stated above, every frame is a key-frame.

8.2.2 IK

Inverse Kinematic (IK) is an animation technique that you specify only the positions of the end effectors. Software calculates all the rotation angles of the middle joints in the joint chain to reach the position of an end effector. If you think of your arm as an IK chain, you grab your wrist and move your arm around by moving the wrist. This is a basic way to think of an IK chain (Figure 8.6). Your shoulder should stay in the same location, but the upper and lower arms will rotate as the hand is moved to different locations.

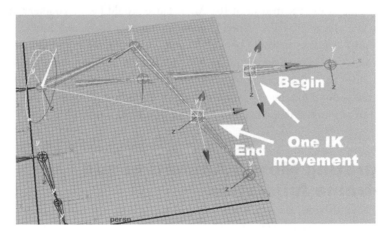

Figure 8.6 *Arm moved by IK*

IK is usually used for the limbs of a body and is usually part of a character rig application. The IK on the arms usually relies on the wrist's position to position the rest of the arm. The ankle's position usually define the position of the leg. If set up properly, the IK chain knows which way to bend, so if the foot is brought up off the floor and close to the body, the knee will flex in the proper direction and the upper and lower leg joints will rotate to accommodate this. The end of the IK chain is usually called an end effector (Figure 8.7). The end effector is the device that's used to position the IK chain. Usually the position of the end effector is key-framed.

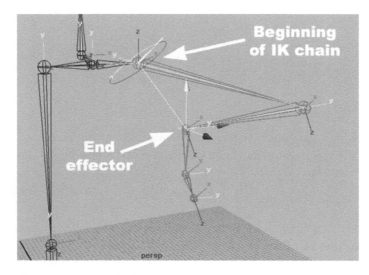

Figure 8.7 *IK end effector*

Sometimes it is easier to use FK instead of IK. An instance would be a character's arms swinging back and forth as the character walks. When you are animating a character's limbs, IK is a powerful tool. If IK weren't available, it would be very hard and time consuming to key-frame a character's walk since you would have to figure out angles of all the joints in the character's legs. With IK, you can prevent a foot from sliding or going through the ground by placing and keeping the foot's end effector on the ground for the appropriate number of frames. A combination of IK and FK gives the maximum amount of flexibility and this usually leads to a better rig.

8.3 Integrating Mocap Animation and Key-frame Animation

Mocap gives us FK key-frames at every single frame of motion. IK allows us to animate entire joint chains without having to specify every single joint rotation. Let's look at what happens when we mix the two.

8.3.1 Why do we want to do that?

Why do we want to mix FK and IK? The answer to the question is simple: to get the best performance possible. Motion capture works great for human motion but people usually conserve motion and therefore do not have an "animated" quality to their actions. Being able to change the motion data is one of the keys to getting what's wanted out of a performance.

Having a way of animating on top of mocap data to add to the actor's performance will give you a perfect method for getting the timing and weight transfer of human motion. The ability to exaggerate motion will put much more life in the character. Motion capture, at its best, is a tool for animators to build on. It is what Disney animators did with rotoscoping for classic Disney animation. They did not simply use the rotoscope to trace over. They pushed the positions further to give even more energy to the characters' motions. Motion capture is not a black box that simply produces data that you are stuck with.

8.3.2 Setting up a skeleton for FK and IK

There are several ways to make custom rigs in Maya and other 3D packages. We are going to concentrate on two methods. One that you will have to put some work into, and a simpler one that Maya gives you for free. Let's start with the more difficult way first.

Switching between FK and IK has been common with regular animation rigs. Back before this was automated for you, there was a special process you'd have to go through in order to use FK or IK data, and this was to use a three skeleton system.

If you are going to use a three skeleton system, the joint orientations and lengths between joints should be identical among the three skeletons. The method is for applying motion to the skeleton by blending FK and IK and not for retargeting motion for differently proportioned skeletons.

Your base skeleton is an initial skeleton that has nothing applied to it, no IK or FK. Since mocap data and IK will be applied to duplicates of the base skeleton make sure that it works with your motion editing or retargeting software and IK tools. Try a few different variations. Once you have the best skeleton for your animation, make two duplicates of the base skeleton. One of these will be for IK and the other will be for FK. The joint rotation angles of these two skeletons will be blended with weights to generate the joint rotation angles of the base skeleton. For blending we will use a Blend Colors utility and a little widget that we will create. It allows us to change all the blending weights for all the joints at once.

Rename all the FK joints to something like "FK_..." and the all IK joints to something like "IK_...." If the three skeletons' joints share names, it can be confusing and cause problems with scripts, constraint, and other methods.

Select the IK, FK, and base skeletons by selecting the top joint in each skeleton's hierarchy. Look at all of the joints in Hypergraph and click the "Input and Output Connections" button (Figure 8.8). You'll see just what goes into and out of each joint. Being able to see connections will be helpful in the next few steps.

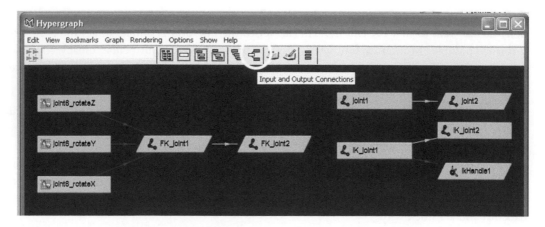

Figure 8.8 *Input and output connections*

Now we need to connect the nodes together in a way that will allow us to blend IK and FK. We want to push the rotation angles of the FK and IK joints into a blender that sends the result out to our base joint. We will do this with a "Blend Colors" utility (Figure 8.9). The Blend Colors utility

Figure 8.9 *Blend Colors utility*

can be found either in the Hypershade or the Create Render Node option under the Rendering drop-down menu in the Hypergraph.

Create a Blend Colors utility. Connect the Rotate output of a joint in the FK skeleton to the first color input (Color1) of the Blend Colors utility using the Connection Editor (Figure 8.10). Connect the Rotate output of the corresponding joint in the IK skeleton to the second color input (Color2) of the Blend Colors utility. Take the output from the Blend Colors utility and connect that to the Rotate input of the corresponding joint in the base skeleton. You need to do this for every joint, and make sure that you always put the FK joint's output in the Blend Colors utility's first color input (Color1) and the IK joint's in the second one (Color2). Make sure it is consistent. (A Maya file with just a simple joint chain set is on the CD.) The reason for using two copies of the base skeleton is so that all the joints are aligned in exactly the same direction. If this is not the case, this setup may not work.

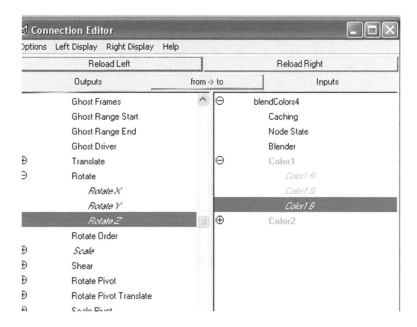

Figure 8.10 *Connecting joint's rotation to Color1 input of Blend Colors*

If you are interested in making this faster, write a MEL script that will run through all the joints and make connections.

We are now going to set up a way to control the FK/IK blend without having to hunt down all the Color Blend utilities. Let's start with the FK/IK control for the right arm. Create a NURBS circle near the right shoulder joint, rename it "rightArmBlend." Select the circle and go to the "Modify" selection menu at the top and choose "Add Attribute." Name it "FKIK Blend." Set it as a float (floating-point) value, give it a minimum value of 0, a maximum value of 1, and a default value of 1. This means the default will be FK. If you let the default value be 0, the default will be IK.

Using Connection Editor, again, connect "rightArmBlend" and one of the Blend Colors utilities associated with the joints in the right arm. Use "FKIK Blend" as the output of "rightArmBlend" and the "Blender" as the input to the Blend Colors. Repeat it with the Blend Colors utilities associated with other joints in the right arm. Once these are connected, changing "FKIK Blend" will change the FK/IK blending of all the joints in the right arm. You need to repeat this with the left arm and both legs.

Now that it's done, we have an FK/IK switcher. It is not an on/off switch but a slide switch. This is useful in all types of situations. We explained all the steps so that you understand how a three skeleton system works. There is also a much easier way. Whenever you create an IK in Maya it will create an IK Blend attribute on the IK handle that allows you to switch back and forth between FK and IK. Thus, with IK Blend you can have mocap data on a skeleton and IK to override it using a single skeleton.

There are other methods as well. Maya has a skeleton system within it now that is very similar to the full-body IK skeleton in MotionBuilder. If you create a skeleton in the way Maya wants and assign the full-body IK to the skeleton, Maya will automatically rig the skeleton. Be careful with using something like this. It has a certain way of doing things and does not allow you to change the rules. It is possible that it will save you quite a lot of rigging time but it is also possible that you can lose some flexibility.

You can also use marker data to directly move IK end effectors (IK handles in Maya) if you set things up correctly. Using marker data and a decent IK setup in Maya, you can create an entire rig and sidestep the rig in MotionBuilder. We will go over that in Chapter 10 where we describe facial rigs with IK and again in Chapter 11 where we talk about puppetry.

8.3.3 Adding key-frame animation to mocap

A very common occurrence in motion capture is that after you have captured quite a bit of data on a generic motion capture skeleton you are reluctant to add any more to the data. We are not talking about editing the data, but adding to what you have or mixing a few different methods together. A common example is adding some finger animation to the hands. Many times the hands are captured as rigid objects that only have three degrees of freedom at the wrists and there is no motion for the fingers. Adding finger animation to this is creating key-frame animation from scratch. You need to rig the fingers and animate them. Pay attention to the attitude in the motion of not only the hand but also the overall body. The observation will give you clues on how fast to animate the hands or how far to move the fingers. Always go back to your reference video from the shoot as well to get ideas about what the fingers were doing.

Another area that is commonly added later as key-frame animation is the facial expressions. Sometimes facial expressions are captured but often they are animated by hand. You create a separate head for facial expressions and generate facial expressions using shape interpolation (shape blend), key-framing, and/or other methods. The animated head has all the facial expressions on it but no rotation or translation. The skeleton with mocap data has rotational values for the neck and head. The animated head can be directly added to the mocap skeleton. However, you may need to tweak the mocaped head motion in order to create the best look for the end product. If you do not, the character may have facial expressions that look out of character or jerky head motions that do not match the facial expressions.

9 Hand Motion Capture

Hand capture is very difficult and at the same time very valuable. Emotions can be conveyed with subtle movements of hands and fingers. When dealing with motion capture, however, fingers give you very small areas to place markers on, which create some interesting issues and challenges.

9.1 Anatomy of a Hand

As with any study of motion capture, anatomy is the best place to start. You need to understand how the hand is structured, how it moves, which type of hand motion you want to capture, and what kind of motion you are willing to sacrifice for efficiency.

Let's start with the forearm, which is the arm between the elbow and the wrist. It supplies the "twisting" of the hand. This rotation around the longitudinal axis of the hand is called the x-axis rotation of the hand. This twisting of the hand is originated in the forearm by the *radius* rotating around the *ulna* (Figure 9.1). However, in animation rigs it is often treated as rotation applied directly to the wrist, having nothing to do with the forearm.

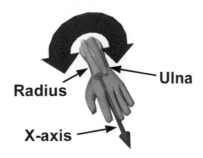

Figure 9.1 *The x-axis rotation of a hand*

The hand rotates around the y- and z-axes (two latitudinal axes) of the hand as well. Let the latitudinal axis that sticks out of the hand from its side be the z-axis of the hand. There is almost 180 degrees of rotation around the axis (Figure 9.2).

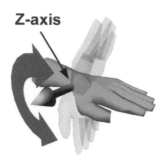

Z-axis

Figure 9.2 *The z-axis rotation of a hand*

The *y*-axis is the other latitudinal axis. It sticks out of the back of the hand. The rotation around the *y*-axis has a smaller range than the rotations around the other two axes (Figure 9.3a). It only rotates about 30 degrees in the adduction (Figure 9.3b) and only 20 degrees in abduction (Figure 9.3c).

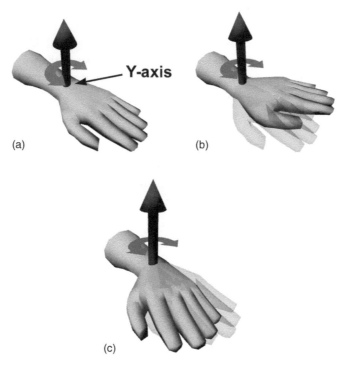

Y-axis

(a)

(b)

(c)

Figure 9.3 *(a) The y-axis rotation of a hand; (b) The y-axis rotation of a hand in the adduction; and (c) The y-axis rotation of a hand in the abduction*

The combination of the three rotations generates the main rotation (i.e., local motion) of the hand, but we may be interested in other motions, such as finger motion and hand cupping. So, let's look at the bones of the hand (Figure 9.4).

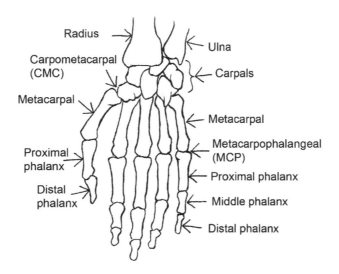

Figure 9.4 *Bones of the hand*

The bones in the hand include several small bones called the *carpal* bones. We have been familiarized with the word "carpal" by carpal tunnel syndrome. (Carpal tunnel syndrome is a medical condition caused by the median nerve compressed at the wrist.) The carpal bones are all different in shape and all have some small amount of motion, but for the sake of simplicity we tend to group them together into one unit. The base of the group of the carpal bones attaches to the radius and the ulna. This is normally where we place the origin of our hand axes.

The *metacarpals* are the longest bones in the hand. These are the bones that allow you to cup your hand. The thumb's metacarpal moves more than the other fingers'. That allows the thumb to rotate and face the other fingers. We use the metacarpals whenever we grip something. The metacarpals are very frequently overlooked in motion capture but are essential if you want to create natural hand motion.

Each finger has three bones called *phalanges*, which are *proximal phalanx*, *middle phalanx*, and *distal phalanx*, except for the thumb. The thumb has proximal and distal phalanges but not middle phalanx. The joint between the proximal phalanx and the metacarpal is called the *metacarpo-phalangeal (MCP)* joint. The MCP is biaxial, in other words, it has two degrees of freedom (DOF) rotation. On the other hand, joints between two phalanges have one DOF rotation (Figure 9.5). As the MCP joint is biaxial, all the fingers can rotate around the two latitudinal axes (y- and z-axes). However, they do not rotate in the longitudinal axis (x-axis) voluntarily. If you place your hand on your desk, you can tap the desk with your fingers by moving them up and down. You can also spread out your fingers, but you can't twist your fingers like your forearm. Fingers and thumbs may be twisted a little when they come in contact with another body, but there are no voluntary muscles to twist them.

Figure 9.5 *Two DOF at MCP and one DOF at joints between phalanges*

The joint between the metacarpal and the carpals is called *carpometacarpal (CMC)*. The thumb's CMC joint is called a "saddle" joint because it looks like a pair of interlocking saddles. The joint is biaxial like the MCP joint.

One important note is that, we have opposable thumbs. Our thumb can oppose the other fingers and allows us to hold things easily and pick up small objects. The thumb does not rotate into the side of the hand but slightly under the hand (Figure 9.6a). Also the thumb is not on the same plane as the fingers. It extends down from the hand at an angle (Figure 9.6b). Relax your hand and pay close attention to how and where the thumb is situated. Setting up the thumb's MCP joint and CMC joint properly is very critical to moving the thumb in a realistic manner.

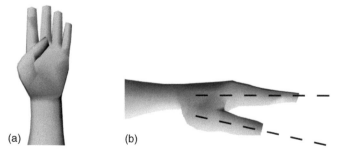

(a) (b)

Figure 9.6 *(a) Thumb rotates under the hand; and (b) Thumb extends down from the hand at an angle*

Important note: Since the thumb is anatomically different from the other fingers it requires different treatments. To avoid potential confusion, a digit refers to any finger, including the thumb, and a finger refers to any finger, excluding the thumb, in the rest of this chapter.

Try to get someone to take top, side, and front photos of your hand in a relaxed pose, in a fist, and with your fingers and thumb spread out. This should give you a pretty good general idea of the range of motion of the hand. Books on anatomy and biomechanics are great references as well.

Now we are more familiar with how the parts of the hand move. Let's look at how to rig a hand for several different variations of hand capture.

9.2 Rig and Marker Set for the Hand

There are several ways to rig a hand for motion capture. Each rigging method calls for a certain number and configuration of markers. We will be looking at rigging methods and marker sets together in this section because we can't talk about one without the other.

All the marker sets for the hand are relatively simple and there are a couple of important criteria that apply to all the hand marker sets. The first one is that all the markers should be placed on the backside of the hand, not on the palm side. Markers are more visible to cameras on the back side. The second one is that markers placed on a finger should be as centered as possible on the finger. If a marker on a finger is too close to one side, it can be easily covered up when the finger gets close to another finger next to it.

9.2.1 Rigid hand

Let us start with the simplest setup for the hand. It requires three "hand markers" or markers on the back of the hand (Figure 9.7) and treats the entire hand as a single rigid piece. The hand motion

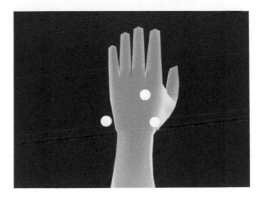

Figure 9.7 *Marker setup for the rigid hand*

is derived from the three hand markers. The hand moves as if all the fingers and thumb were bound together and could not move at all.

All you need from your mocap data is the wrist joint's rotation. Apply the rotation to your rig's wrist to rotate the entire hand. If you do not have rotational or skeletal data, use the translations of the three hand markers. Create point and aim constraints to have the three hand markers rotate the hand (Figure 9.8). (See Section 6.2.3 for how to orient an object using three markers and point and aim constraints. The same method can be applied here.)

When you are attaching three markers to a hand, create an irregular triangle with the markers, but not a regular triangle that has three edges of the same length (Figure 9.7). Varied distances among the markers make it easier for you and the tracking algorithm in your mocap system to identify the markers. Add one or two extra markers. They will help you deal with occlusion problems.

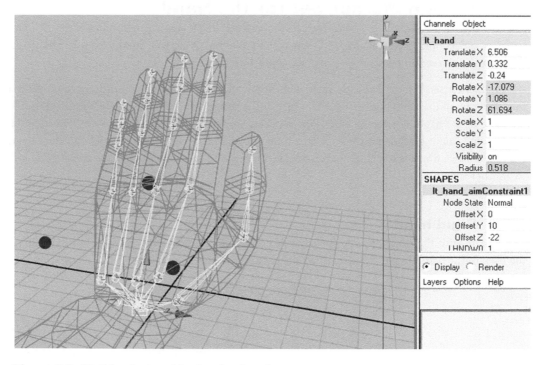

Figure 9.8 *Rigid hand oriented by three hand markers*

This method is probably the most commonly used one and useful for many types of projects. However, your character will lack details in hand motion that would convey meanings, emotions, intents, directions, and other communications. Ask yourself if capturing the most basic hand motion is enough for your project. If not, try one of the following methods.

9.2.2 Mitten

The second method is what we call the mitten. A marker on a fingertip, or a "fingertip marker," is added to the rigid hand's marker set and the fingertip marker is used to drive the motion of all the fingers and thumb (Figure 9.9). Since there is just one marker on the selected finger, usually the middle finger, and no marker on the other fingers, the rig basically curls all the fingers together to reach the marker. With this method you have the hand motion (motion of the upper part of the hand) and the finger motion (motion that is shared by all fingers and thumb).

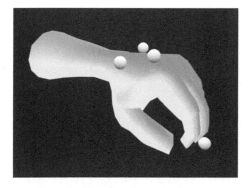

Figure 9.9 *Marker setup for the "mitten"*

The rig needs a chain of joints to control the finger motion. Create a chain of four joints. The first joint should be the MCP joint, the second one at the joint between the proximal and middle phalanges, the third one at the joint between the middle and distal phalanges, and the fourth one at the

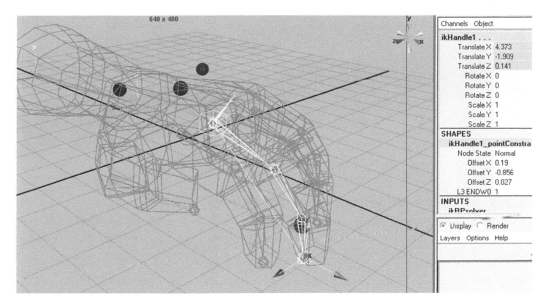

Figure 9.10 *Joint chain in the middle finger*

fingertips. The rig can have either a single joint chain shared by all the fingers or a joint chain in each finger. In the former case, the skeleton of the hand looks anatomically wrong but works fine if the skin of all the fingers is bound to the single joint chain appropriately. In the latter case, each finger's skin should be bound to the corresponding joint chain and the joint rotation should be shared across all the joint chains via a script, an expression, or any other method that you choose (Figure 9.10). Either way, motion looks like the motion of a hand wearing a mitten.

Use either constraints or inverse kinematic (IK) to let the finger's joint chain be driven by the finger-tip marker. If you choose to use IK, have the position of the IK end effector (IK handle in Maya) constrained by the position of the fingertip marker.

Copying the finger rotation to the thumb creates a grasping motion (Figure 9.11). An expression or script is a good tool for copying the rotation since it can easily scale the finger's rotation angle to fit the range of the thumb's rotation angle.

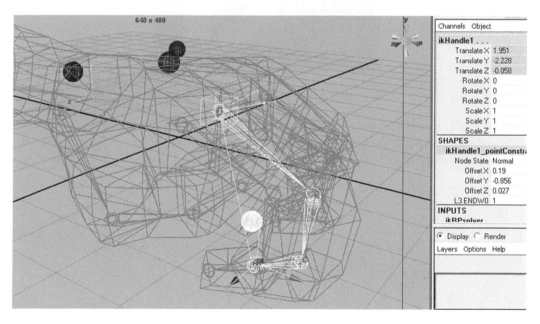

Figure 9.11 *Grasping*

9.2.3 Mitten with an independent thumb

A variation of the "mitten" method adds a marker on the thumb or a "thumb marker," to the basic mitten's marker set. The thumb marker, not the fingertip marker, is used to drive the thumb's motion (Figure 9.12).

When you add a joint chain for the thumb to your rig, recall the characteristics of the thumb, for example, the thumb has two (not three) phalanges, more movement in its metacarpal than other digits, the thumb's MCP joint is offset from the upper hand, and the thumb rotates in a different

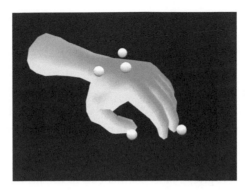

Figure 9.12 *Thumb marker added to the basic mitten marker*

direction from the other digits. Keeping that in mind, create the first joint of the chain at the CMC joint, the second one at the MCP joint, the third one at the joint between the proximal and distal phalanges, and the fourth one at the fingertips. The thumb's joint chain has four joints like the middle finger's joint chain but anatomically the joint placement is different (Figure 9.13).

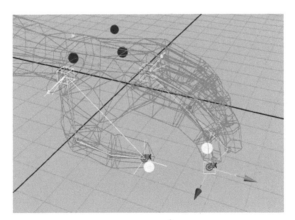

Figure 9.13 *Markers on the thumb and middle finger*

Use either constraints or IK to let the thumb's joint chain be driven by the thumb marker in the same way as the finger's joint chain is driven by the fingertip marker.

If higher DOF than necessary is given to the thumb, the thumb tends to acquire weird rotations. So give one DOF rotation to each thumb joint, except for the first joint. The first joint with two DOF rotation allows the thumb to move to desired locations.

When you are placing markers on the middle finger and thumb, place them on the distal phalanges but leave a small distance from the fingertips. The small distances prevent markers from being occluded and being mistaken for one big marker when the digits get close to each other.

You can get more with this approach than the basic mitten method but there are still a lot of limitations. For instance, you can't spread fingers because one finger marker is providing motion for all four fingers.

9.2.4 Mitten that stretches

Another variation of the basic mitten method replaces the fingertip marker in the marker set of the mitten with an independent thumb with two markers, "index finger marker" and "little finger marker."

This configuration of markers gives a fairly natural curl to the hand and allows the fingers to fan out. The index finger marker provides all the motion for the index finger, 67% of the motion for the middle finger, and 33% of the motion for the ring finger. The little finger marker provides all the motion for the little finger, 67% of the motion for the ring finger, and 33% of the motion for the middle finger. In other words, the motions of the index and little fingers that have markers are averaged with weights to generate motions of the middle and ring fingers that do not have markers (Figure 9.14). The thumb marker provides the thumb motion independently.

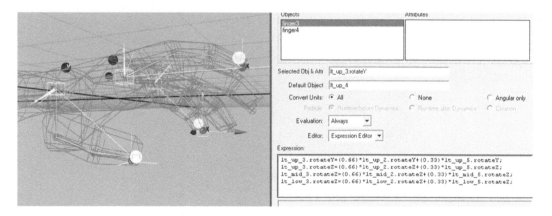

Figure 9.14 *Weighted averages of rotations*

When you are attaching markers to the index and little fingers and the thumb, remember to place them on the distal phalanges but leave small distances from the fingertips. The markers should be closer to the fingertips than the third joints but not too close to the fingertips.

9.2.5 Ultimate

For the ultimate in full hand and finger capture, place markers on all five digits' joints and fingernails, and at least three markers on the back of the hand (Figure 9.15).

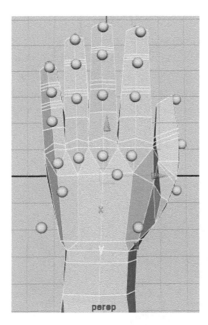

Figure 9.15 *Markers on all finger joints, fingertips, and back of the hand*

When you use this many markers, make sure that the real hand (the hand that you mocap) and your 3D model of a hand (the hand that you apply mocap data to) are similar because the markers drive motion of the fingers, thumb, and metatarsals individually. All major bones of the hand, including the metacarpals, should be represented as joints in your rig.

The MCP joints of all five digits should have two DOF (or three DOF in rare cases). Usually rotations around the longitudinal axes of the fingers and thumb are ignored. Thus, we assume that none of the digits "twists" at all. However, our fingers and thumbs can be twisted by external forces. Your project might require you to accurately represent the twist. If the fingers and thumb need to twist, give three DOF rotation to the joints to the MCP joints; otherwise, leave them with two DOF rotation.

The three markers on the back of the hand, "hand markers," can rotate the whole hand with three DOF as explained in Section 9.2.1. With skeletal or rotational data and a skeleton for the whole body, the location of the hand is normally computed using the rotation angles of all the joints that are above the wrist joint in the full-body skeleton. The hand markers can translate the hand to wherever it should be in the world space if neither skeletal nor rotational data is available or if the hand rig does not belong to a full-body skeleton.

One of the hand markers should be placed on the fifth (i.e., little finger's) metacarpal fairly close to the knuckle (i.e., MCP joint) (Figure 9.16). The reason to have the fifth metacarpal marker close to the knuckle is that the metacarpal moves more near the MCP joint and less near the CMC joint. You want the marker to move as much as possible in order to capture the fifth metacarpal motion, which is an important part of the cupping motion. Your rig should be set up so that the rotation of the fifth CMC joint is shared with the fourth (i.e., ring finger's) CMC joint since the fourth metacarpal moves slightly in a manner similar to the fifth metacarpal. It is important that two DOF rotation is given to the first (i.e., thumb's) CMC joint and one DOF rotation to the fourth and fifth CMC joints. The movement of the second and third metacarpals is insignificant and can be ignored. No DOF is necessary for the second and third CMC joints.

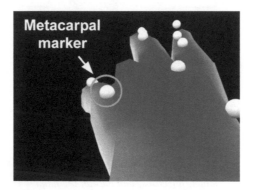

Figure 9.16 *Metacarpal marker*

The marker placements of the four fingers should be identical to each other. The thumb's marker placement superficially looks similar to that of the four fingers but anatomically different as discussed in Section 9.2.3.

The biggest issue with placing "joint markers," or markers on joints, is that the skin over knuckles travels noticeably and a marker placed on top of a knuckle slides when the digit is flexed. So, the markers for the MCP joints should be placed behind the knuckles but not on the knuckles (Figure 9.17). The markers for the second and third joints can be placed on or close to the joints. A fingertip marker should be placed on each digit's fingertip as well.

Be aware that aim constraints can cause backward bending in some digits if not enough attention is given to the orientations and locations of the joints' rotation axes. A little bit of backward bending is not bad since many people can bend their fingers and thumbs backward slightly but a lot of bending looks strange. Look at the hand rig on the CD. Since "up" vectors, that would control where the joint rotation's y-axis points at, are not set up in the rig when the markers move too much, the fingers can "flip" their rotations.

When you have a fully articulated hand rig, make sure that the rotation axes of all the finger joints line up so that the z-rotation is about the same axis and in the same direction for all the

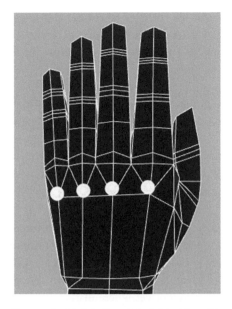

Figure 9.17 *Markers right behind knuckles*

finger joints. (Remember the thumb is different. Set up the thumb's joints accordingly.) The reason why you want to orient the rotation axes of all the finger joints identically is that when you select finger joints and rotate them about the z-axis, you want all of them to curl in the same direction.

In summary, two critical factors to a success in full hand and finger capture with the setup presented above are giving the minimum DOF to all the joints and orienting the joint rotation axes properly.

9.3 Capturing Hands

When you capture hand motion with some type of finger motion you usually have specific reasons to do so, such as your character has to interact with a prop, it is important to show the tension in a character's hand when it makes fists, or maybe you are researching finger motion and you want to see if you can replicate the subtle movements of the individual fingers. No matter what your reason is you have to think about the same set of basic issues that we will discuss in this section.

The biggest issue is visibility. How visible are all the markers to the cameras? How many markers will be occluded if a character makes a fist or two characters shake hands? There are a lot of concerns related to visibility and it can get complicated very quickly. The majority of visibility issues comes from the nature of hand/finger motion itself. A good example is trying to capture someone pointing a finger at something. It is a very simple and universal hand gesture, but when the three fingers fold into the palm of the hand the fingertip markers on these fingers disappear. There will be other markers that are harder for the cameras to see if the thumb is curled down over the three fingers. There is no way to capture the finger pointing gesture without occlusion.

We suggest that you capture all the shots you need, including the ones you know will give you much less than ideal data. And then find out if you can fix the messy data later. You might decide that key-framing the shots is more practical than trying to fix the data. If only one or two out of twenty or thirty shots have serious problems and all the other shots are fine, there is no good reason for you to give up on hand capture. Just use an alternative method for the shots with mocap data that is too difficult or too time consuming to fix.

Explain to your talent what happens if hands are hidden and markers are covered up, and ask them to try not to create actions that will hide hands or markers. At the same time, you don't want your talent to become too conscious of the markers. You really want them to get used to having markers on their hands and act as if markers were a part of them. We caution against stopping a performance because you see one or two places where markers are occluded. Unless it is a very long performance, let it go and finish the shot. Then ask the talent to try it again. Suggest different hand or finger placements and explain why you'd like to make the change.

Marker size is another one of the major factors that determines how much data you will have and how reliable it will be. When you are performing the full hand and finger capture, use the smallest markers possible to get the best separation. This type of capture is often done with a close-up mocap setup (similar to the facial capture setup that we will see in the next chapter). However, mocap systems are getting better every year. There are already several systems that can capture hands, fingers, and full body all at the same time. In the near future non-marker systems may be able to capture full body, hands, and facial all at once in a reasonably sized space in which capture subjects can move around freely.

Think again about the actions that you want to capture and check if anything will get in the way of the markers. If you want to capture someone's index finger pulling a rifle's trigger, make sure that the markers on the index finger do not get caught in the trigger guard. If they do, remove the trigger guard. Prevent preventable issues as much as you can.

Motion capture is supposed to give you speed and accuracy, but you may want to rethink your method if you have to spend months cleaning your data. You may be better off reshooting or key-framing some (or all) of the shots. Knowing the limit of your system helps you when it is time for you to make decisions.

In the next chapter we will look at facial capture. It presents a set of challenges. Some are similar to those we saw in this chapter. Others are unique and different.

10 Facial Motion Capture

The face is the only part of the human body where some voluntary muscles attach to other muscles instead of bones. Most of our facial muscles are small, thin, layered, and embedded in fatty tissue. Motion capture of facial expressions is almost exclusively done with optical systems due to the subtlety of the motion. Small markers of 2 or 3 mm diameters are pasted on a capture subject's face with hypo-allergenic glue used for false eyelashes or wigs. If a high-end system has a sufficient number of high-resolution cameras, it can capture full-body motion and facial expressions simultaneously in a capture volume that is large enough to allow the subject to move around. However, with most optical systems, facial expressions are captured separately from body motion in a smaller capture volume created by placing cameras around a seated capture subject (Figure 10.1). Facial data is "stabilized" by removing the head motion from the data to isolate the local displacement of the facial skin caused by the facial muscles underneath the skin. The stabilized data is applied to a facial rig. Before we examine facial motion capture, let's look at the anatomy of the face to understand the forms and functions of the bones and muscles in the face.

10.1 Anatomy of a Face

Our most important irreplaceable organ, the brain, and the sensory organs for all of our five senses (eyes for sight, ears for hearing, tongue for taste, nose for smell, and skin for touch) are in our heads. The human skull consists of the 8 cranial bones and 14 facial bones and its shape underlies the appearance of the individual. The total of the 22 skull bones forms multiple cavities; the largest one to house the brain and smaller ones to house the eyes, ears, nose, and mouth.

The *cranium* is the dome-shaped part of the skull that contains the brain. It is formed by eight plates of bones. The plates of the unborn baby are not fused so that the head can deform when it goes through the narrow birth canal. If you touch the top of a newborn's head, you can feel a soft spot. There is another smaller soft spot toward the back of the head. The smaller one closes in a few months after the birth; the larger one remains soft until about 2 years of age. The *frontal cranial* bone is the bone that forms the top part of the face above the eyes and plays the most important role in the facial features among the cranial bones. *Arcus superciliaris*, or brow ridge, is a ridge beneath the eyebrows and a part of the frontal cranial bone. Great apes have more prominent *arcus superciliaris* than humans, and men more than women.

The lower front part of the skull, the face, consists of 14 facial bones. The *maxilla*, or upper jaw bone, is two bones fused together. During pregnancy, two bones in the embryo's cheeks grow

Figure 10.1 *Camera setting for facial capture*

toward the center of the face. They meet at the center under the nose and form the maxilla. If they fuse incompletely or fail to meet, the child is born with a cleft palate. Because our diet consists of mostly processed and/or cooked foods, our food is much softer than what our ancestors ate. Our jaws have grown smaller than our ancestors'. That is the reason why we have wisdom teeth, molars for which our jaws don't have space for. Men have larger jaws than women but it seems that men suffer from wisdom teeth as much as women do. The facial bones that are most important for the facial features are the *mandible* (lower jaw bone), maxilla, *zygomatic* (cheek bones), and *nasal* bone (Figure 10.2).

The *temporalis* muscle covers the large temporal area of the face. It starts from a side of the cranium, goes under the *zygomatic arch*, and attaches to the top of the mandible. Great apes and monkeys have larger temporalis muscles than ours. Their temporalis often covers the entire sides of the cranium and reaches the crest that runs from the *arcus superciliaris* to the back of the cranium. Our smaller temporalis muscles are another example of devolution that the human has gone through as we evolve. The *masseter* muscle starts from the *zygomatic arch* and attaches to the lower part of the mandible. There are other muscles that are involved in chewing and moving the tongue, in addition to the temporalis and masseter. The muscles related to chewing and swallowing food are located in the

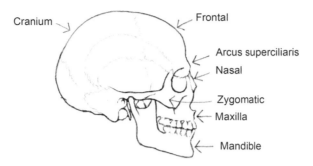

Figure 10.2 *Human skull*

sides of the skull and the throat area. A note here is that our mandibles move in all three directions. We can move our mandibles up and down, front and back, and right and left. But we often simplify and restrict the movement of a 3D character's lower jaw by allowing the jaw joint to have only one degree of freedom (rotation around a horizontal axis).

Facial muscles cover the front part of the cranium, facial bones, fat, cartilage, and other types of tissues in the face in layers. Combinations of these muscles create facial expressions. The total number of facial muscles may differ from one anatomy book to another because some of them function together and are hard to separate. The *orbicularis oris* muscle is the circular muscle around the mouth. All the other muscles around the mouth are used to open the mouth, while the *orbicularis oris* closes the mouth. We use the *orbicularis oris* to blow a whistle or a brass instrument. Adjacent to the *orbicularis oris*, the *buccinator* muscle is thin and rectangular and covers a relatively large part of the cheek. The buccinator tenses the cheek and pulls the corner of the mouth and flattens the cheek (Figure 10.3).

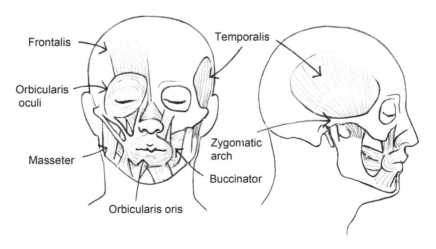

Figure 10.3 *Facial muscles*

The *orbicularis oculi* is the circular muscle around the eye. Similar to the *orbicularis oris*, the *orbicularis oculi* is the only muscle that closes the eye. The *frontalis* is on the frontal cranial and is not anatomically classified as a facial muscle. However, it is the muscle that raises eyebrows and is important in facial expressions. Botulinum toxin, a toxic protein, is injected into this muscle as a cosmetic treatment to temporarily smooth wrinkles on the forehead.

There are a number of anatomy books for artists. We suggest that you buy an anatomy book if you don't have one yet. It will be beneficial when you design a marker set and a rig for your 3D character. In addition to anatomy, morphology and morphogenesis of the face are useful. Form follows function. Every part of our faces (and bodies) has the shape determined by the functions that it has or used to have. Knowing how all these different skeletal and muscular systems work with each other will only help you create better outcomes.

10.2 Camera Setup and Capture

You can't paste many markers of a standard size (e.g., markers with the diameter of half an inch) on a human face. Even if you succeeded in it, markers would be too close to each other and the mocap cameras would consider multiple markers as a single giant marker. So, we need to use smaller markers and higher resolution for facial capture. If your system consists of a good number of high-resolution cameras, you may be able to do facial capture without removing the cameras from the walls or reducing the capture volume. If you are not sure, test it. If your cameras cannot track small markers, you will need to bring the cameras down, adjust the camera lenses' focuses, and possibly replace the lenses to allow for focusing on closer objects.

Place a chair to seat a capture subject and the cameras on tripods around the chair (Figure 10.1). There is no need for cameras to look at the back of the capture subject's head but make sure that some cameras are covering the sides of the subject. The coverage of 180 degrees to 200 degrees around the subject allows the subject to rotate the head from side to side to some degree. If the subject turns the head too much to one side, some of the markers won't be seen by any cameras. For instance, if a subject turns his head to his right by 90 degrees, the markers on his right temple won't be seen by any of the cameras. You need to talk to your capture subject about it before you start capturing.

The cameras should be placed at various heights. Place one third to one half of the cameras a little bit above the eye level of the seated capture subject. Place the other cameras below the eye level and let them look upward at the subject. The passive optical system's cameras emit bright lights and being surrounded by them is like being interrogated – it is hot, glaring, and uncomfortable. It is a good idea to place your cameras so that there is a void of cameras in the middle right in front of where the subject will be sitting. This gives the subject someplace to focus his/her eyes that isn't covered by red lights and is also a convenient place to add a video camera to take reference video.

It is wise to check the camera setup and calibrate the system before asking a capture subject to take the seat. Instead of the capture subject, you can use an object of an approximately human head size (e.g., a football). Paste markers on the object and place it roughly at the seated subject's head position (Figure 10.4). The system should be calibrated with a smaller wand with smaller markers instead

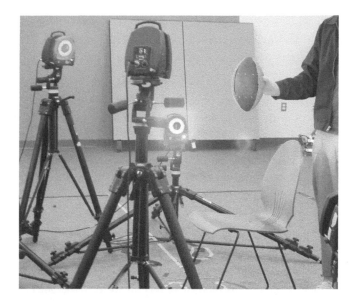

Figure 10.4 *Checking camera setup with a markered football*

of a regular-sized wand that you use for full-body capture. And remember you will be behind the cameras and lights, and the subject won't be able to see you. Have a rehearsal with the cameras off or at a different location with a similar seating arrangement before starting capture sessions.

Have a facial capture kit with small markers, hypo-allergenic glue, tweezers, hair bands, and pins. Small markers are easier to handle with tweezers. A hair band and pins can keep the capture subject's hair off the face. Ask your capture subject to wear a top in a dark color on the capture day since bright-colored clothes (e.g., a white T-shirt) reflect light. If your capture subject comes in a light-colored cloth or if you want markers on the chest, you can use a mocap suit. As in full-body capture, it is a good idea to take reference video when capturing facial data.

10.3 Facial Rig

A facial rig for facial motion captured data can be set up in a number of ways. We will introduce a few methods here. Try all and find which method works best to achieve the goal of your project.

10.3.1 Facial rig with discrete joints

The simplest way is probably attaching markers to a set of joints that are bound to the facial skin geometry. To set up a rig with this method first create a set of discrete joints which do not have hierarchical relations to each other (whereas the joints in a skeleton are normally in a hierarchy). The locations of the joints on a 3D character's face should be similar to that of the markers on the capture subject's face. Next, bind the skin geometry to the joints. Unless you are animating a robot's face, which consists of rigid parts with no flexible part whatsoever, you need to use a skin binding method that allows each vertex of the skin geometry to be influenced by multiple joints in order to have smooth deformation over the skin. Finally let the position of each joint be constrained by that of the corresponding marker with a distance offset.

The main issue of this method is that controlling influences or "painting weights" for each joint tends to be a time-consuming tedious labor. You need to determine which joints (i.e., markers) influence which vertices of the skin geometry and how much as you check how the skin geometry is deformed by mocap data.

If you are a Maya user, the skin binding method that you want to use for this method is smooth bind, not rigid bind. Rigid bind divides the vertices of the skin geometry into clusters and attaches a cluster to each joint. As the result rigid bind allows each vertex of the skin geometry to be influenced by a single joint. You can decide which vertices a particular joint moves by changing the membership of the joint's cluster. Rigid bind is quicker to set up than smooth bind, but a crease appears on the skin where two clusters meet, justifying the name of rigid bind. On the other hand, smooth binding allows each vertex of the skin geometry to be influenced by multiple joints with weights.

When you use a smooth binding method, it is very likely that the automatic default weighting method initially binds many joints (markers) to vertices in an undesirable manner. Painting weights in a graphical way allows you to change the weights of vertex points and reassign vertices to appropriate joints on a joint by joint basis. This task is easier if all the joints have the names that let you identify them quickly. This is a very good reason to name your joints in a logical descriptive manner.

Unfortunately painting weights with many vertex points can sometimes reassign some weights to joints that are far away from the vertex points that you are working on and that should not influence the points. When all else fails, flood all the vertex points to just one joint, then use the component editor to weight each point by hand, then smooth all of these out a little. It's time consuming, but usually gives better overall results than trying to fix the weights that went out of control.

10.3.2 Facial rig with muscles

Another approach is to create "muscles" that are polygons placed under the facial skin around the mouth and eyes and on cheeks, forehead, chin, and any other parts of the face you want to be animated by mocap data. You deform the skin geometry indirectly by attaching markers to the vertices of the muscles and using the muscles as influence objects. Figure 10.5a shows a skin geometry and Figure 10.5b shows "muscle" polygons for the skin.

To set up this rig in Maya, first create a single joint which can be anywhere in the facial model and bind the skin geometry to the joint using smooth bind. Secondly create "muscle" polygons in such a way that the locations of the polygon vertices mimic that of the markers on the capture subject. Triangulate the muscle polygons and let them be the skin geometry's influence objects. Thirdly turn on the Use Components mode which you can find in the skin geometry's skinCluster node and paint weights for the influence objects. Finally create a cluster at each vertex of the muscle polygons and point constrain a marker to the corresponding cluster with the offset position on. Note that the Use Component mode is off by default and if it is off the skin geometry will not be deformed by mocap data.

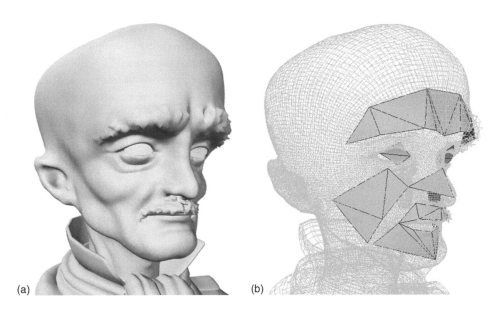

(a)　　　　　　　　　　　　　　(b)

Figure 10.5 *(a) Skin; and (b) Muscles*

With this method you need to paint weights for smooth binding as required for the rig with discrete joints but weights are painted for each muscle, not for each joint, and painting weights is much less time consuming.

10.3.3 Facial rig with IK

As we saw in Chapter 8, if your 3D animation application allows you to blend inverse kinematics (IK) and forward kinematics (FK), you can take advantage of it. In case of animation of a full body (or partial body), rotational mocap data applied to a skeleton is essentially an FK animation for us. We can modify it (e.g., retarget a limb) using IK. By blending FK and IK we can connect a mocap driven animation segment and a key-frame animation segment seamlessly as well. If you plan to manipulate facial expressions after mocap data is applied to a rig or desire to combine key-frame animation and motion capture data driven animation, having a skeleton with IK in the face of a 3D character is useful. However, we suggest the use of IK for mocap data driven animation and FK for key-framing. That is a reverse of what we do with a rig for body mocap data.

The reason for the reversed rolls of IK and FK stems from the fact that the facial muscles move only the soft tissues but no bones, except for the ones that move the lower jaw, while voluntary skeletal muscles in the rest of our bodies are designed to move our skeletons. Mocap applications provide us with a template human skeleton with bone segments that represent major bones in our bodies, such as spines, femurs, tibias, and humeruses. It is a gross approximation of a human skeleton but we can agree on a relatively simple skeleton as an effective template; 40–50 markers are enough to capture the skeletal movement of a single capture subject. Mocap applications compute rotation angles for each joint in the skeleton.

On the other hand, with facial animation the rig can be as simple as or as complicated as an individual project requires. Similarly we may use just one dozen markers to a couple of hundred markers depending on the subtlety of the expressions needed for the animation. There is no standardized skeleton for the face. Therefore, mocap applications do not provide us with a facial rig that has rotational data associated with joints. Optical mocap data only gives us positional data. Thus, facial mocap animation is not FK animation. So, how can we blend it with key-frame animation? We let positional mocap data drive IK and use FK for key-framing. It is difficult to animate limbs using FK. Therefore, we use IK for that, but it is easy to key-frame a facial rig with FK. Look at Figure 10.6. The upper eyelid "skin" joint and the lower eyelid "skin" joint are placed on the surface of the skin geometry. The upper eyelid "muscle" joint and the lower eyelid "muscle" joint are placed in the interior of the skin geometry. Since the eyelids slide on the eyeball as the eye is opened and closed, placing the eyelid "muscle" joints at the center of the eyeball deforms the eyelids in a realistic manner when the eyelid muscle joints are rotated.

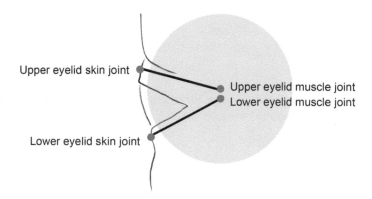

Figure 10.6 *Eye IK*

The first step of setting up a facial rig with IK is creating "skin" joints and "muscle" joints. Place your fingers in front of your ears and open and close your mouth. You can feel where your lower jaw bone meets your cranium. A "skin" joint for the lower jaw movement can be placed at the center of the chin while the "muscle" joint for the lower jaw should be between the two ears at an equal distance from each ear. Similarly, for other parts of the face place "skin" joints on the surface of the facial model and corresponding "muscle" joints in the interior of the model. The locations of skin joints should approximate the positions of markers. If the distances between pairs of skin and muscle joints are fixed for markers on lips, your 3D character's lips cannot protrude. Inserting a middle joint between a skin joint and a muscle joint allows the distance between the skin and muscle joints to be changed. Middle joints are useful for the mouth and cheeks.

The next step is setting up IK: create an IK from a muscle joint to the corresponding skin joint and attach a marker to drive the IK. (In Maya, that is point constraining a marker to an IK handle with a distance offset.) When key-framing is required, turn off IK, and change and key the rotation angles of the "muscle" joints. Aligning all the "muscle" joints' local coordinate systems with the world coordinate system or orienting the "muscle" joints local coordinate systems so that they rotate in a desired manner will simplify your key-framing.

With this method you need to paint weights for each skin joint. Thus, painting weights can be as tedious as for the rig with discrete joints in Section 10.3.1 but you have a control mechanism (i.e., a skeleton with FK and IK) to manipulate facial expressions driven by mocap data.

10.4 Marker Set

As mentioned in the previous section, the number of markers necessary for facial capture widely varies from project to project. There are a few rules, though. One is that you should not place more markers than you need on a capture subject's face. With full-body motion capture, extra markers can be used to compute rotation angles and lessen the effect of marker occlusions on the computation. Extra markers also may be used to form a rigid body to fill in missing data. However, with facial capture, data generated by extra markers will be disregarded. If two adjacent markers are close to each other they may be seen as one marker by some cameras. So, if you know one of them is not necessary, remove it.

Another rule is to have two kinds of markers: one set of markers to track the displacements of the facial skin by facial expressions and speech and another set of markers to track the movements of the head. The latter may be named "stable" or "static" markers, although skin slides in every part of the human face or head, to some degree. Place your hands on your head, move your eyebrows up and down, and open and close your mouth. You can feel your muscles and skin moving under your fingers. The "stable" markers are placed on the parts of the face that are moved the least by facial expressions or speech, such as along the hairline (or on a hair band) and between the eyes. Place a fingertip between your eyes and repeat squinting while looking for a spot where the skin does not move much. If you move your fingertip up or down by just a couple of millimeters, you can feel much larger displacement of the skin by squinting. See Figure 10.7 for a sample facial marker set. Among the 38 markers, 5 are stable markers.

Markers on the upper eyelids are the hardest ones to deal with. Both of this book's authors have eyes that are difficult for facial mocap. Brian's brow is high and his upper eyelids can get hidden under his eyebrows. I have "single eyelids" with no crease on the upper eyelids; common among Asians, whereas most Caucasians and African people have "double eyelids." Markers placed close to my eyelashes disappear when I open my eyes. Markers placed a few millimeters from my eyelashes do not move much. Although my son has double eyelids, the creases on his upper eyelids are very close to his eyelashes, giving him more of an Asian appearance than Caucasian (Figure 10.7). When I tried to capture my son's facial expressions, I could not place markers on his upper eyelids effectively because markers on his upper eyelids disappeared when his eyes were open, just like on me. The parts of his upper eyelids above the creases do not move much nor the markers placed there.

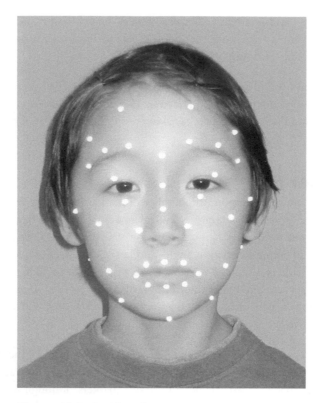

Figure 10.7 *Facial marker set*

If you need to capture a subject who has eyes like my son's or mine, and if the subject agrees, you may paste markers on the eyelashes. (My son didn't agree... .) It may take a few minutes until the capture subject gets used to the feeling of the markers moving in such a close-up way and stops blinking. If your capture subject has eyes with typical double eyelids like my husband's (Figure 10.8), place a marker between the eyelash and crease on the upper eyelid to get the maximum movement of the upper eyelids. Notice that the markers on the upper and lower eyelids are placed in such a way that they keep a distance even when the eye is closed. If they get too close to each other when the eye closes, they will be seen as one marker and marker swapping and mislabeling problems can happen.

Remember that when you capture a face, you capture the movements of soft tissues. On the other hand when you capture a full body, you capture the skeletal movement. There is a layer of fat between the facial skin and skull. Also there are fat pods under eyes, in cheeks, and chins. The thicknesses of fat pods vary depending on the race, age, gender, and other conditions. These fat pods provide facial mocap data with secondary motions. The secondary motion is subtle but it can give a remarkable realism to mocap data driven facial animation. Design a marker set that provides you the data you need for your project and find a capture subject who has the most suitable facial features and the ability to act.

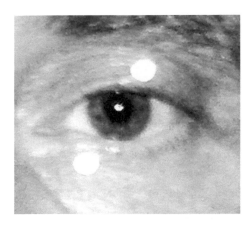

Figure 10.8 *Markers on eyelids*

10.5 Facial Data Stabilization

One way to get rid of the head movement is to not allow the capture subject to move the head in the first place. To keep a capture subject from moving the head, you can use a chair with a head support and tie the subject's head to the head support or use a head brace that is designed to keep a patient with a neck injury from moving the head. However, it is natural for us to move our heads while we speak and express our feelings. In fact, head movement is an essential part of our non-verbal communication. So, a better approach is to allow the capture subject to move the head while capturing facial data. The head movement can be removed from facial mocap data using "stable" markers discussed in the previous section with relatively simple mathematics.

The Facial Data Stabilizer (FDS) is a script written in MEL (Maya Embedded Language) and is on this book's CD. MEL is similar to the C language and if you are not a Maya user we hope FDS will be rewritten in the scripting or programming language of your choice without difficulty. FDS considers head motion, which we want to eliminate, as the rotation and translation of an object in the world space while the deformation of facial features, that we want to keep, as translation of the object's vertices in the object (local) space.

The usage of FDS follows:

1. Select stable markers in the right temple area.
2. Select stable markers in the left temple area that have the mirrored positions of the right temple markers selected in Step 1.
3. Select stable markers in the upper forehead area that are on or very close to the line that divides the face into the right and left halves and/or pairs of markers where the markers of each pair have mirrored positions in the right and left halves of the face.
4. Select stable markers in the lower forehead area that are pairs of markers where the markers of each pair have mirrored positions in the right and left halves of the face.
5. Select all the markers to be processed.
6. Stabilize.

The main mathematical ideas used in developing the FDS's algorithm are the following.

A right-handed local coordinate system (local space) with respect to the world coordinate system (world space) is formed as follows:

- *x*-axis: a vector from the averaged position of the stable markers in the right side of the face to that of the left side. After the *y*- and *z*-axes are computed the *x*-axis is recomputed as the cross product of the *y*- and *z*-axes to ensure that *x*-, *y*-, and *z*-axes are perpendicular to each other.
- *y*-axis: a vector from the averaged position of the stable markers in the lower forehead area to that of the upper forehead area.
- *z*-axis: the cross product of the *x*- and *y*-axes.
- *Local origin*: the averaged position of the stable markers.

The relationship between the coordinates of a point in the local space and the coordinates of the same point with respect to the world space is defined as follows:

$P * A + \text{local origin} = P'$, where

$P = [x, y, z]$ are the coordinates of a point in the local space
A is a 3 × 3 matrix such that
      1st row of A is the *x*-axis of the local space
      2nd row of A is the *y*-axis of the local space
      3rd row of A is the *z*-axis of the local space

$P' = [x', y', z']$ are the coordinates of the point in the world space

We know the marker positions in the world space but not in the local space (i.e., P' is known and P is unknown for us). As described above, the matrix A and *local origin* can be computed. Thus, they are known as well.

By moving *local origin* to the right side, the above equation becomes:

$$P * A = P' - \text{local origin}$$

By multiplying both sides by a matrix B, an inversion of A:

$$P * A * B = (P' - \text{local origin}) * B$$

Since $A * B$ is an identity matrix I

$$P * I = (P' - \text{local origin}) * B$$

Thus,

$$P = (P' - \text{local origin}) * B.$$

Hence, by finding a matrix B that is an inverse matrix of A we can compute the marker positions in the local space. Since the matrix A's rows (the axes of the local space) are perpendicular to each other, the determinant of A is never 0. A is never singular and A's inverse always exists. So, a simple algebraic method for inverting a 3×3 matrix works fine here.

Let M be a matrix A at the current frame when the script is executed (or the frame the user selects). $P * M$ gives us P'', the coordinates of a marker in the world space without the head movement. Note that A and B are different for each frame but M stays the same for all frames in the sequence.

$$P'' = P * M = (P' - \textit{local origin}) * B * M$$

Let $N = B * M$ then

$$P'' = (P' - \textit{local origin}) * N$$

Thus, the coordinates of the markers in the world space without the head movement can be computed by one 3×3 matrix multiplication per marker.

If mocap data has large head movements and/or "stable" markers are not as stable as we want them to be, FDS leaves some head movement in the processed data. If that happens, apply FDS again to the processed data. Applying FDS to the data two or three times should remove head movement sufficiently.

The Head Movement Isolator (HMI) is another MEL script on the CD. It isolates the head movement using stable markers on the subject's chest area. The HMI's algorithm is identical to the FDS's. The data stabilized by HMI can be used to add the head movement back to the mocap driven facial animation. To use HMI, remember to place at least four markers on the capture subject's chest area. The chest markers should form a rectangle (or a regular shape with more than four edges) and should not all line up on a single straight line.

If you need facial animation and full-body animation both driven by mocap data and if your system can capture both at the same time, do so. If you need to combine facial animation driven by facial data and animation driven by full-body data, the head movement can come from either facial data or body data. When the 3D character to which the motion is applied is far from the camera (e.g., in a long shot) then the facial expressions will not be too readable for the audience and facial expressions and head movement do not have to match. (In such cases facial expressions are not crucial to have in the first place anyway.) So, you can use the head movement from the body data. When the character is in a close-up or medium shot, you want the head movement to match the facial expressions. So, it is better to use the head movement from the facial data than from the body data. When the character is not too close to the camera or too far from the camera (e.g., in a wide shot) it is your call.

For a Maya-based facial animation project that required the head motion driven by mocap data in addition to animations of facial expressions and speech by mocap data and key-framing, I used both FDS and HMI. I set up two skeletons for the face. One skeleton has skin joints, muscle joints, and FK/IK as described in Section 10.3.3 for the facial expressions. The other skeleton is much simpler than the first one. It has just a few neck joints for the head movement. For facial animation the skin was bound to the FK/IK skeleton by smooth bind and the skeleton was animated by the mocap data stabilized by FDS. For the head motion a lattice deformer of the skin was bound to the simpler skeleton by rigid bind and the skeleton was animated by the mocap data stabilized by HMI. In short by using the smooth bind and indirect rigid bind on a lattice, the same skin geometry was animated by two types of mocap data: one that had facial animation and the other that had head movement.

See the frames from two animation sequences: the top one was created using FDS, and the bottom one was created using FDS and HMI (Figure 10.9). You can see that the animation with head motion is more expressive than the one without.

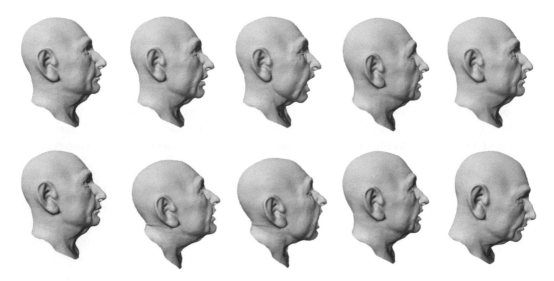

Figure 10.9 *Facial expressions without head movement (top row) and with head movement (bottom row)*

10.6 Facial Data Editing

Our faces are not like Mr. Potato Head. Our facial expressions are displacements of our facial features that can't switch places. You can raise your hand above your shoulder or lower it below your knees but no matter how hard you move your nose it stays above your mouth and below your eyes (unless you stand on your hands). Especially after stabilizing facial data the x-, y-, and z-translations of facial markers stay within relatively small ranges. By looking at an upper eyelid marker's y-translation we can tell when the capture subject blinked or squinted and when the eye was wide open. By looking at the y-translations of lip markers, we can tell when the capture subject opened and closed the mouth. The x-translations of markers at the mouth corners can tell us when the capture subject smiled or pouted. Of course, we can tell when a foot of a full-body capture subject was on the floor or not by looking at the y-translation of a marker on the foot. However, it is not easy to figure out what's going on in the x- and z-translations of the marker since the capture subject could have been moving in any direction. We can say facial mocap data behaves in a more predictable way.

The above fact makes it easier to edit facial data than full-body data. When a facial marker's data goes out of its normal range you can tell that the marker was occluded and/or mislabeled. For instance, when you find a frame range in which the y-translation value of the right upper eyelid marker is higher than the lower bound of the y-translation of the right eyebrow marker, you can tell there is a problem. Simply lowering the y-translation value or copying the y-translation value of the left upper eyelid marker for the frame range can fix the problem.

Let's say the y-translation value of a marker is too low (or too high) for a frame range. If you delete the keys for the marker's y-translation for the frame range and let a spline interpolation fill in the gap, then the subtle jittering that is a characteristic of mocap data will be lost. You could move each key but that is a time-consuming task if the frame range is not short. Lattice Deform Keys Tool in Maya's Graph Editor is a useful tool to edit facial data. It is one of the Transformation Tools and you can find the tools under Edit in the Graph Editor. Lattice Deform Keys Tool allows you to move multiple keys quickly without losing the jittering (Figure 10.10).

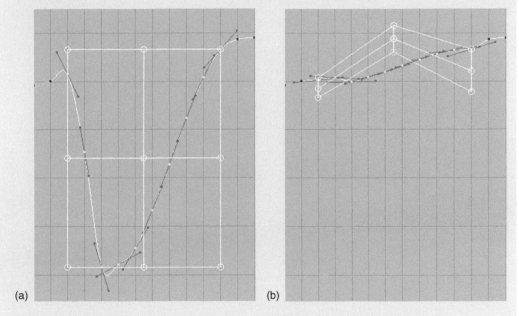

(a)   (b)

Figure 10.10 *Before (a) and after (b) keys are moved by Lattice Deform Keys Tool*

No matter what you're looking to do, or which approach you want to take, make sure that the ends justify the means. If you only need a few points to get corners of the mouth and eyebrows, then only use markers for that and not a full facial rig. If you need an extremely expressive and precise face, go all out with the markers and don't rely on any interpolation to make up for gaps between markers. Think facial capture through before you start using it and you'll get better results. Facial capture can be very frustrating to work with, but with adequate preproduction and some testing you'll get very convincing results.

11 Puppetry Capture

Motion capture of puppets, or "puppetry capture," is capturing a puppet's movement that a puppeteer creates. Markers are attached to the puppet and the puppeteer's motion is captured indirectly.

Physical puppets that you motion capture do not have to be real puppets. They can be anything that you can attach markers to. Digital characters that you apply mocap data to can be anything too. Anything from computer generated counterparts of real puppets to set pieces, props, shape-shifting geometries, changing colors and lights, etc.

In puppetry capture, a puppeteer manipulates a puppet with his/her hand. This motion may become a digital character's foot or eye or something else. You have to think about your end character and what you are trying to capture for the character, not about a puppeteer's arm being the arm of a character.

Think about how you can construct a motion for a digital character using the motions from different parts of a puppet (and a puppeteer) and how a puppeteer can manipulate these parts. There are all kinds of tricks and original ways to create complex digital characters; you just have to start thinking in a new way.

Puppetry capture merges motion capture and puppetry art to create a unique hybrid. This may let you see motion capture in a different light.

11.1 Background

I spent approximately 8 years doing all kinds of motion capture professionally. There were a lot of video games and a handful of film projects that were all exciting to work on. However, for the most part, it was all about the same. Every couple of weeks I captured someone walking, running, shooting, throwing, etc. After a while, they all started to merge together. It was around that time I asked myself "What else can I do with these little balls of reflective tape?"

I grew up with the Muppets and always enjoyed puppets, especially the ones created by Jim Henson. After watching behind the scenes on movies with live action puppets (such as the "Dark Crystal" and "Labyrinth") I realized that there were many more interesting ways to take a human form and create something totally different. I wanted to create all kinds of creatures and merge them with motion capture.

Since then I have created simple marionettes, made puppets out of foam, attached rigs to my body, mimicked full-body puppets, and captured a folding chair. The idea is to take rather simple objects and drive them with complex human manipulation. The overall effect is something unique and different, and at times very useful and charming.

Puppetry capture is not new. One example would be where people have captured the motion of prosthetics and magnetic markers embedded in puppets. What's been missing is a comprehensive overview of the methodology of puppetry capture, which will enable motion capture to become a much more creative medium than simply a tool that seems to be dedicated to tracking a person's movement and applying it to a digital character.

Most of the puppetry capture works I have done to date (some of which will be discussed in Section 11.5) are along the lines of prototypes and proof of concepts, but I hope they will encourage others to explore new innovative ways to use motion capture. In the next few sections I will try to highlight several of the characteristics of puppetry capture.

11.2 Benefits

One of the benefits of puppetry capture is that you don't need to retarget the puppets at all because you can build a physical puppet to the exact proportion of a 3D puppet (or vice versa). This helps tremendously in post-processing. The data from puppetry capture usually needs very little cleanup. It is normally limited to the areas where there is occlusion. The data from puppetry capture can be attached almost immediately to a 3D puppet skipping the intermediate steps that human mocap data goes through.

Interaction between characters normally requires a fair amount of post-processing to look right for the characters and scene. Interaction between puppets requires very little attention. If you create a pair of physical puppets with two different sizes, A and B, and a pair of digital puppets with sizes, C and D, so that A:B = C:D, you will not encounter retargeting issues common for multiple subject capture.

Puppets have a good amount of secondary action or what I call "physics for free." A secondary action is one of the principles of animations and is a direct result of another motion. For instance, when a person walks his/her hair swings. Secondary actions add realism to animation. When you produce a key-frame animation you work in a 3D space with no physical rules. You have to create everything from scratch, including secondary actions and other aspects of motion caused by the laws of physics. Animators need to painstakingly key-frame or use a dynamic simulator to add realism to the actions of a character and the motions of everything else that moves in a scene. On the other hand, it is inherent for puppetry capture to catch the laws of physics in action.

Another advantage is that puppets never have to change, for example, puppets do not need to take a bathroom break, which would necessitate a human capture subject's mocap suit to come off and on and move the markers attached to the suit. Thus, the markers on a puppet stay where they are supposed to be. There are no shifted or lost markers that you experience with human capture.

You can have just one physical puppet and have many different puppeteers perform with it for a number of 3D puppets. The first time the puppet is calibrated is the last time unless some markers are knocked off or added/removed. A lot of content can be rapidly created with just one puppet.

Puppets can be big time-savers since they are always there and ready to be captured. Also there are puppeteers in almost any town. Skilled ones can show you amusing puppetry techniques that you may have not imagined and can be added to your production.

Puppeteers can help you create characters with items from local dollar stores. You would be surprised how easy it is to create fairly complex creatures from everyday objects. If you have a larger budget or access to a machine shop, then there is little limit to what you can produce; however, it does not really matter how the physical puppet looks like since the 3D puppet is the only one that people will see. It always helps to put eyes on the puppet so the puppeteer knows where the puppet is looking.

During a performance a lot of improvisation and unscripted interaction among performers often-times happen. The nature of live action performance can generate an extremely large amount of content in a short amount of time. This only works if the pipeline for physical puppets and 3D puppets has been established in preproduction. A few experiments beforehand will create a smoother pipeline.

11.3 Ideas/Inspiration

Traditional puppetry works are often inspirational. For instance, one idea came from Big Bird to use a hand to operate the head and mouth of a character. A broken hierarchy (Figure 11.1) was created for

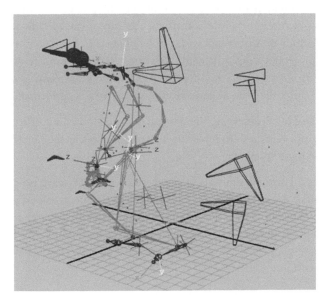

Figure 11.1 *Full-body broken rig setup*

the entire skeleton and used a regular marker set to move all of the different skeletal parts. A broken hierarchy is a segmented hierarchy that consists of joint chains that can be moved independently.

Another idea is capturing a puppeteer's motion directly and indirectly simultaneously to drive certain aspects of a digital character. Markers are placed on a puppeteer's body and the props (e.g., PVC pipes) attached to the puppeteer. Both the puppeteer's motion and the puppet's motion (props' motion, in this case) are captured at once. The puppeteer's motion and the props' motion can be used as the motion of different limbs of a character, a parameter for shape interpolation, the color of a digital character, or anything else. The combinations are as endless as your imagination.

Once I worked on an idea about capturing a human hand alone and creating creatures with the hand data, so I used two of my students, Eric Camper and Josh Huber, to help with creating and capturing the motion. There was a good amount of data with an interesting quality quickly generated and I only used a small amount of it. The end results were imaginary creatures that resembled jellyfish and a crab and that ran around.

I recommend watching behind the scenes on live action puppetry films, such as the "Dark Crystal" and "Labyrinth". See how puppets were created and how puppeteers used innovative techniques. These films help you design a marker setup that suits your creature and more importantly they will fuel your imagination.

11.4 Performance

There is a direct connection between the puppeteer and the puppet. They can give very complex human motion to simple objects. The puppeteer's motion creates an illusion of life in the puppet. In a recent conversation with master puppeteer Phillip Huber, he offered the following comments: "The good thing about motion capture is that you also capture all the idiosyncratic actions. That has been my argument for the direct link with a good puppeteer, because it is essential that the soul of the puppeteer is projected clearly through the strings, sticks, or even remote controls. I am not sure why or how this works, I only know what I feel when I see good performance that comes from the heart. I do think the little imperfections help to bring a performance closer to reality."

Motion capture is capable of preserving the spontaneity and all the little things that are so hard and time consuming for other methods to replicate. The performance is the key to any motion capture, whether based on acting or puppetry.

There have been times I created a puppet and performed with it having no clear idea on what the digital character would look like. The majority of these experiments ended with poor results. The best performance comes out of puppetry capture when a digital character is at least sketched out prior to capture and the puppeteer has some idea about how it will look. Show a 3D model of your digital character to the puppeteer, if you have it modeled already. If not, have some sketches. It is hard to describe it with just words.

Puppetry capture allows you to think about a puppeteer's movements and how they influence the digital character in different dimensions. For example, you can generate a lot of unique ideas with just legs. The feet are in connection with the floor, but why do the legs need to be connected to the feet? Who's to say how long or short the legs need to be? If you make the legs shorter, you now have the rest of the legs to use to influence motion. If you decide to make the legs longer, you could forego a torso for your character and have the legs connect directly to the head.

You can break all of your physical puppet's motion into parts. One puppeteer can be responsible for the head of a digital character, another the legs, etc. You can have body parts flying all over the place if you want.

Generally a single puppeteer is best to create the performance for one digital character. Set up a collection of objects and devices so that a single puppeteer can manipulate and create the entire motion for a character by herself/himself. Occasionally a character needs to be operated by multiple puppeteers but it is difficult to synchronize everyone, especially if there is any improvisation.

11.5 Projects

This section has brief descriptions of some of the puppetry capture projects that I worked on over the years. The prime goal of the projects was to prove that they could be accomplished, they were repeatable, and they could be quickly set up after some initial testing.

In one puppetry capture I asked two people to be in different parts of the mocap space, (myself and a student, Fran Kalel), to provide motion for one character. One moved a chair with markers on it while holding a stick with three markers in front. The chair became the body of the creature floating through the air, and the stick became the creature's giant eyeball (Figure 11.2a).

The other was in a corner of the capture space moving a "markered" prop from side to side in a corner of the capture space. This caused the giant eyeball to rotate in its socket. The second person occasionally waved a single marker vertically in the air. That motion drove the blinking of the eye (Figure 11.2b).

(a)

Figure 11.2 *(a) Chair and rod puppet; and (b) Markers controlling eye*

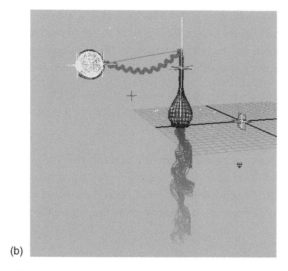

(b)

Figure 11.2 *Continued*

The data was written out of the mocap system with no cleanup other than filling in data gaps caused by occlusion. See the movie file in this chapter's folder on the CD. You can see some of the jitteriness of the character in the movie. The data was not filtered because the unfiltered motion worked better for the character.

The goal for the next project was to mimic what a real hand puppet could do using a bare hand. Several markers were attached to a hand directly. Initially there were too many markers. The number of markers was reduced down to the minimum necessary to make a digital character move as if it were manipulated by a puppeteer via a real hand puppet. It seemed that simply replicating a real puppet's performance could not justify the time and effort put into the project unless something different could be done with it. As the result, two characters flying around in a cosmos were operated by one puppeteer (Figure 11.3).

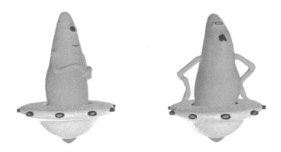

Figure 11.3 *Two characters manipulated by one puppeteer*

After that project, I created a puppet with multiple controls. Overall movement of the puppet was controlled by one of my hands, the eyes by my feet, and the eyebrows by my other hand (Figure 11.4). It took a little while for me to get use to coordinating everything, but once I had it figured out, it worked well.

Figure 11.4 *One puppet*
with multiple controls

In another project, I placed markers on my waist and feet, on the water bottles attached to the PVC pipe that was hanging off of me, and on a couple of flashlights in front of me. The dynamics of the water bottles swinging around created interesting motion (Figure 11.5).

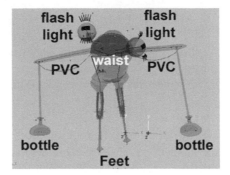

Figure 11.5 *Odd placement of markers*
for a creature

11.6 Methods

There are several methods for applying puppetry capture data to a 3D character. One is to use constraints. For instance, a combination of point constraint and aim constraint can be used to translate and rotate the root joint of the 3D character and the IK end effectors of the character's limbs.

Let's suppose that you are creating constraints for the root joint of a digital character in Maya. Create two transform nodes (empty groups). Parent one of the empty nodes under the other and parent the root joint under the two transform nodes so that the joint is at the bottom of the hierarchy. You will be using the top transform node for translation and the node below for rotation (Figure 11.6). In this way, if you decide to apply stretch and squash to the character, you can insert another transform node between the two existing nodes and apply scaling to the new middle node while ensuring that rotation, scaling, and transformation are applied to the character in that order. (Read the next chapter for the order of transformation.)

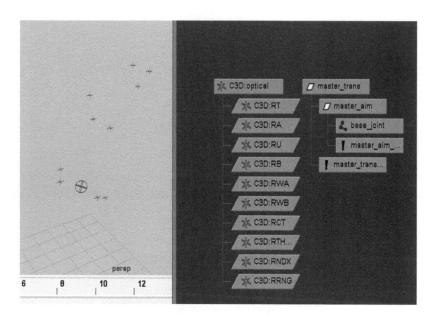

Figure 11.6 *Setting up position and aim for a joint using mocap markers*

Select a marker that will provide the location of the root joint. Let's call it the base marker. Point constrain the base marker to the translation node. Select another marker that the x-axis of the root will point toward. Let's call it the aim marker. Select one other marker that the y-axis of the root will point toward. Let's call it the up-vector marker. Aim constrain the aim marker to the rotation node using "Object Up" option as "World Up Type" and the up-vector marker as "World Up Object" (Figure 11.7). Here, you are using three markers to orient the root joint of the character, which orient the whole character's skeleton. This is identical to how a prop is oriented using three markers as seen in Chapter 6.

When you use this method, make sure the local rotation axes of the transform nodes and the joint are oriented identically. Having them oriented in the same way makes it easier to determine if the aim constraint is working or not. This is because in Maya you can see joints but not transform

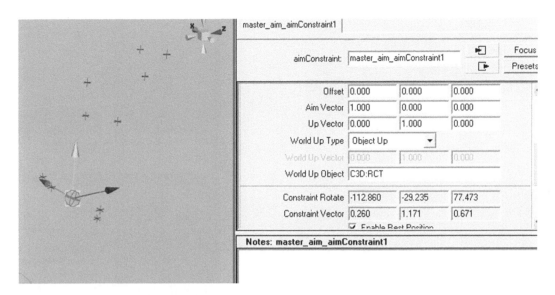

Figure 11.7 *Aim constraint with Object Up option*

nodes. You can turn on the option to display local rotation axes, but since the two nodes and joint are all stacked on top of each other in a 3D space, you can't see their rotation axes.

If you do not use "Object Up" option as "World Up Type" in the aim constraint, the character may not rotate as you expect. Let's suppose that two markers, instead of one marker, are used as aim markers for the aim constraint and that the two markers are rotating around the stationary point that is at the shortest equal distance away from both markers. As the two markers rotate, the root joint (and the character) should rotate around the x-axis of the root joint. However, without "Object Up" option, the character won't rotate because the x-axis of the root joint will keep pointing at the point between the two aim markers while the y-axis of the root joint will keep pointing upward in the world space. If "Object Up" option is selected as "World Up Type" and a marker that rotates around with the two aim vectors is selected as "World Up Object," the y-axis (and z-axis) of the root joint will rotate around the x-axis of the root joint, that is, the character will rotate around the root joint's x-axis.

For the arms and legs, create IK. Point constrain appropriate markers to the IK handles (Figure 11.8). You may want to place markers on the character's elbows and knees or wherever you want the limbs to bend. In the rig, let the limbs bend toward the markers. (In Maya pole vector constraint does that.)

Try to keep your rig as simple as possible. Usually the simplest solutions are the most effective and least likely to break.

The broken rig allows each segment of a character to be an independent object. One example, the hips and spine in one hierarchy; the head and neck in another; each arm, leg, hand, and foot in its own. A broken hierarchy can be used effectively with unfiltered marker data. The combination of

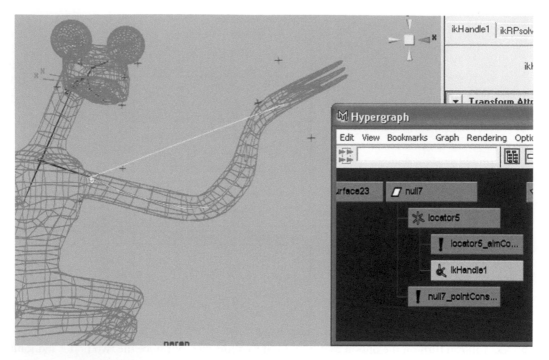

Figure 11.8 *IK chain moved by mocap markers*

a broken hierarchy and unfiltered marker data preserves a lot of secondary action and weight/balance transfer that tend to be lost when marker data is fit into a skeleton. Since there is some stretching and pulling between the segments of a character with a broken hierarchy it has a different look from a character with a normal connected hierarchy.

11.7 Real Time

Real-time feedback is very important for puppetry capture. Jenny Stoessner whose Ph.D. dissertation is on puppetry researched the role of puppetry in TV. Through several conversations, she informed me that puppeteers seeing their creations in the monitors, even though the motion was reversed from their perspective, played a very important role in the quality of their performances. Being able to see how the digital character acts and what is around the character on the screen as you perform is essential.

When you look into puppetry, especially shows such as *Sesame Street* and the *Muppet Show*, you can see how puppeteers give lifes to the puppets. When a new puppet is created, a puppeteer works with it until the puppeteer finds a voice, movements, and personality for the puppet. There have been times when a puppet doesn't work well with a puppeteer, another puppeteer takes it, and a whole new puppet character, voice, and actions come out. There is a special connection between the puppet and the puppeteer and having a real-time feedback is essential.

There are some drawbacks with real-time puppetry capture. One is that an initial setup may take time. That means the puppeteer needs to be on set for a long time for the first time or two but with good preproduction that will work out.

Another disadvantage of real-time capture is that when the mocap equipment goes down either because of hardware or software glitches, it may or may not come back on in a timely manner. This can be a cause for alarm for the production staff. So, do your best to minimize all production difficulties. Make sure that puppeteers are not wearing light reflective clothing or shoes. Tell them how large the capture volume is, so they will not take the puppet out of the space. Make some commonsense judgments and always test the equipment out before performers arrive.

A drawback, that is getting to be less of an issue these days, is the look of real-time digital characters. We used to use low-resolution character models for real-time puppetry capture. Low-resolution characters did not look as nice as characters in non-real-time applications. This is still somewhat true, but the graphics cards are getting so powerful that the divide between real time and non-real time is getting smaller. Puppeteers tend to have a pretty good imagination, so let them work with low-resolution characters if your graphics card can't handle real-time rendering of complex models.

If you use a bluescreen you can place digital characters and real actors on a virtual set. Have the puppeteers wear blue. Actors, puppets, and props can have any colors except blue on them. They are filmed in front of a bluescreen. By keying blue out, the puppeteers and the background (i.e., the bluescreen) become transparent and only the digital puppets, actors, and props are composed over the virtual set. This is all possible in real-time and can broaden the capabilities for content creation. However, be careful with the logistics when setting this up. There are a number of elements that need to come together seamlessly to make this work.

The right arm of a hand puppet (or a human capture subject) does not always have to move the right arm of a digital character. Try indirect manipulations with puppets and props. Make abstract characters and think of ways to have the characters move and emote. Have a character manipulated by multiple people. Try real-time capture.

There will be projects for that the above information is useful. There will be times when a combination of performers and markers creates a life form that is no longer confined to a body with two arms and two legs but something unique is created, but some characters can be better animated by key-framing or procedurally. Find and use the best approach to create the world that matches the vision of the project.

12 MoCap Data and Math

In this chapter we will look at mocap data generation, data types and formats commonly used in mocap, and mathematical definitions and concepts that frequently appear in mocap and other areas of 3D computer graphics and animation. Mathematical topics covered in this chapter are coordinates and coordinate systems, order of transformation, Euler angles, gimbal lock, and quaternions.

Artists can take full advantage of motion capture technology without knowing how mocap data is stored in a file format or mathematics involved in the technology. However, the information presented in this chapter will help you understand commonly used jargon that you might have already encountered in software manuals, professional web sites, and conversation with programmers. That understanding should let you get the results that you want with less time spent on trial and error. We believe that the knowledge that you gain in this chapter will be especially useful when you are setting up and troubleshooting a production pipeline or when you face a technical problem while working on a project.

Mathematics is like a foreign language. It has its own vocabulary, symbols, and grammar (rules). Mathematical notations are designed to represent matters in our world as mathematical equations in a precise and abstract matter. For those who are fluent in the language of mathematics, mathematical equations are a beautifully convenient way to express ideas and communicate with others. However, for those who are not mathematically trained, most of mathematics beyond arithmetic is incomprehensible. Mathematical equations make many creative people say "It's Greek to me!" So, we will try to explain mathematical concepts with the least number of equations.

12.1 How Data Is Created

Before we look at types and formats of mocap data or mathematics, let us look at how mocap data is created. As we looked at Chapter 1, there are three popular types of mocap systems that are used and each type generates data in a different manner.

12.1.1 Optical systems

Optical systems generate data in several steps. First the cameras in an optical system capture a sequence of grayscale images (or black and white images) of the lights reflected or emitted by markers on a performer. Image processing methods are applied to the images to minimize the noise and separate the markers from the background. To determine the center of each marker, circles are fitted

179

to all the markers (and items other than markers, such as reflective materials on sneakers in a camera view if you are not careful); 2D camera data is created by determining the 2D coordinates of the center of each circle for each camera view. Sometimes instead of using circle fitting, a centroid calculation is performed to determine the center of the marker. The centroid calculation can be used with circle fitting to obtain a better estimation of the position of the marker's center, resulting in better 3D data. This is because markers in the distance and partially occluded markers often appear as irregularly shaped blocks of pixels, far from circular shapes, in camera views. Using the 2D camera data, camera coordinates, and triangulation of the 3D coordinates of each marker are computed and the trajectory of the marker throughout the captured sequence is identified in a 3D coordinate system.

At this point the data has sequential 3D coordinates (moving positions) of each marker but no rotational information. This translational data may be used for facial animation with no further processing. When the translational data is fitted to a skeleton (a set of joints in a hierarchy), translational and rotational data is computed for the root joint (the top joint of a skeletal hierarchy, sometimes also referred to as the hips) and rotational data (but no translational data) is computed for all the other joints. Such data is called skeletal data. A skeleton driven by skeletal data is essentially a forward kinematics animation. Thus, when an animation sequence of the skeleton is rendered, the position of a joint (other than the root joint) will be computed using the position of the root joint and the rotation angles of all the joints above the joint in the hierarchy. In some systems, translation or scaling along the axis of the back and neck is also calculated to represent the compression and tension of the spine.

12.1.2 Magnetic systems

Data generation is more straightforward with magnetic systems than optical systems since both the positions and orientations of tracking sensors placed on a performer are captured. Three mutually orthogonal sets of coils in the transmitter source or "box" produce an electromagnetic field of three frequencies that fills the tracking space. The tracking sensors must be in this field for detection. Three orthogonal coils in each tracking sensor generate a resonant response and the system detects the orientations as well as the positions of sensors. Magnetic systems are sometimes called "six degrees of freedom systems" because x-, y-, and z-translations and x-, y-, and z-rotations are captured. One word of caution with a magnetic field is that one side of the field is "up" and the other is reversed, so if the data appears upside-down, have your mocap subject move to the other side of the field and the data will be right side up.

12.1.3 Mechanical systems

Mechanical systems measure joint rotations using potentiometers or similar measurement devices. The relative positions of joints are computed using the rotational information. Global translation is not measured unless an electromagnetic sensor or some other positioning device (e.g., an inertial sensor) is added. Recall that optical systems capture positions and compute rotation angles. Mechanical systems are opposite: they capture rotation angles and compute positions. Some systems will also measure how far someone has walked by estimating it with footsteps. Because the feet need to be in contact with the ground, you won't see data for motions like jumping or rolling on the ground.

12.2 Data Types and Formats

There are two main types of motion capture data: translational data and rotational data. As we saw above, due to how data is generated, translational data is native to optical systems; rotational data to mechanical systems; both to magnetic systems. Translational data is useful for facial animation, puppeteering, and real-time applications. Rotational data is useful for full-body animation with a skeleton. In the rest of this section we will look at how data is stored in a few popular formats.

The C3D, ASF/AMC, BVH, and FBX formats are probably the most commonly used formats for motion capture data. C3D, ASF/AMC, and BVH are the formats that have been developed specifically for motion capture, whereas FBX is for 3D animation in general. The ASF/AMC and BVH formats store hierarchical skeleton data, while the C3D format does not. Let's look at them individually.

12.2.1 C3D

C3D is a binary file format for motion capture data used in animation, biomechanics, and gait analysis to store motion capture data. The format is flexible enough to store 3D coordinates and any numeric data in a single file. For instance, it can store 3D data and analog data in a synchronized manner. The C3D format is composed of a number of 512-byte blocks of information, where this 512-byte block size is simply an artifact originated from the early development of the format in the Fortran programming language.

The *header section* is the first block in the C3D file. The header allows software applications to quickly obtain information about the file content without reading the next section (parameter section) which has a more complex structure for the rest of the file. The first word in the header points to the start of the parameter section in the file. The header contains other information, such as the number of trajectories, the number of analog channels, the trajectory sample rate, and the analog sample rate.

The *parameter section* normally starts at the second block in the C3D file. The length of the parameter section varies. This section contains the information necessary to read the data contained in the file, such as the name, data type, dimension, and location of each *parameter*, where parameters are descriptions of types of the data stored in the file, but not actual data.

The *data section* contains actual data which can be 3D data, 2D data, analog data, or any combination. The C3D format is in the public domain and the detailed information necessary for implementations with the format can be found at www.c3d.org.

12.2.2 ASF/AMC

ASF (Acclaim Skeleton File) and AMC (Acclaim Motion Capture) are files of the mocap data format developed by Acclaim Entertainment, Inc. Acclaim was a video game company, one of the key contributors to the development of optical motion capture technology, but went out of business in 2004. Acclaim invented a unique procedure for generating skeletal data from optical marker data along with this file format. After the format went into the public domain, Vicon adopted it as its output format.

The two files that the format consists of are: *ASF* and *AMC*. The paired file format was developed due to the fact that the same skeleton is often used for many different motions. A skeleton is stored in a single file, instead of multiple motion files. One advantage of the ASF/AMC format (and the BVH format) over other file formats is that each joint can have its own order of transformations while in other formats the same order of transformation is applied to all the joint segments. Another advantage of the ASF/AMC format is that the files contain both global rotational data and hierarchical (local) rotational data of all the segments.

The *ASF* file contains all the information of the skeleton, such as its units, documentation, root node information, bone definitions, degrees of freedom, limits, hierarchy definition, and file names of skin geometries, but not the data itself. In addition, the ASF file contains the initial pose to which all data in the associated AMC files is relative. The ASF file is divided into sections and each section starts with a colon and a keyword (e.g., :units, :documentation, :skin, and :root).

Note that the ASF file format does not allow a child segment to attach to its parent segment with an offset. As a result, when a gap is needed between a parent segment and a child segment, you need to create an additional segment with no degree of freedom between them to fill the gap. Another restriction of the ASF file format is that only one root can exist in a file. Thus, if two skeletons are needed, you need to attach them to the shared root.

The *AMC* file contains the actual motion data for the skeleton defined by an ASF file. The bone data is sequenced in the order as the order of transformation specified in the ASF file.

12.2.3 BVH

The BioVision Hierarchical (*BVH*) file format was originally developed by BioVision, a motion capture service company which specialized in sports analysis and animation and is no longer in business. The BVH format is more widely used than the *BVA* format, that is, BioVision's earlier and simpler format. The BVA format is an ASCII file that contains no skeleton hierarchy but only the motion data with a fixed order of transformation. The *BVH* format is a binary file that contains both a skeleton and motion capture data and allows each segment of a skeleton to have a specified order of transformation. A drawback of the BVH format is that it lacks a full definition of the initial pose. Another drawback is that the format has only translational offsets of children segments from their parents. No rotational offset can be defined. Moreover, the BVH format is often implemented differently in different applications, that is, one BVH format that works well in one application may not be interpreted in another. All the same the format is very flexible and it is relatively easy to edit BVH files.

The BVH format has two sections: the hierarchy section and the motion section. The hierarchy section contains the definition of a joint hierarchy within nested braces like source code written in the C programming language. Each joint in a hierarchy has an offset field and a channels field. The offset field stores initial offset values for each joint with respect to its parent joint. The channels field defines which "channels" of transformation (translation and/or rotation) exist for the joint in the motion data section of the file. The channels field also defines the order of transformation. A channel is either x-, y-, or z-translation or local x-, y-, or z-rotation. All the segments are assumed to be rigid and scaling is not available.

The motion section of the file contains the total number of frames in the animation, the frame speed in frames-per-second, and a numeric entry for each channel in the same order as in the channels field in the hierarchy section.

12.2.4 FBX

While the data formats we have looked at, C3D, ASF/AMC, and BVH, were developed specifically for motion capture, FBX was not. *FBX* was originally developed by Kaydara for its 3D animation package "FilmBox," which later became "MotionBuilder." Kaydara was acquired by Alias in 2004 and Alias was acquired by Autodesk in 2006. The FBX format is designed to describe animation scenes and is supported by many 3D animation software packages to transfer files among them.

The FBX format can contain geometries, textures, cameras, lights, markers, skeleton, and animation. The FBX file is either ASCII or binary and the entire file is formatted by nested braces like C code. The file is divided into sections and each section starts with a keyword and a colon (e.g., FBXHeaderExtension:, Definitions:, Relations:, Connections:, and Takes:). Each section has subsections that start with keywords and colons as well. The topological relations between segments in a hierarchy are stored in the Connections section and translational data (marker data) is stored in the Relations section. Rotational data (rotation angles of the joints in a skeleton) is stored as animation keys in the Takes section. The FBX software development kit that allows software developers to transfer files into the FBX format is available at Autodesk's web site.

One large advantage that MotionBuilder has over other 3D animation packages is that it can take any of the other file formats (ASF/AMC, BVH, or C3D) and translate them into the .fbx format. This allows MotionBuilder to work as a type of "universal translator" between not only different animation systems, but different types of skeletal structures.

In the rest of this chapter let's look at mathematical definitions and concepts that are useful and important for motion capture and many other areas of 3D computer graphics and animation.

12.3 Coordinates and Coordinate Systems

Coordinates are an "ordered" set of values which specify a location relative to an origin. The word "ordered" is important and emphasized for a reason. Normally with mathematical sets, in what order set members appear in the list of members does not matter. For instance, a set {Roo, Max, Madeline} is the same as {Max, Madeline, Roo}. However, with coordinates the order of the members (values) is important. Sets that have the same values in different orders, for example, (2, 4) and (4, 2), specify different locations. Going two blocks east and four blocks south is different from going four blocks east and two blocks south.

A coordinate system is a system that defines how coordinates are assigned to each point location in the space. Two coordinates that specify locations in two different ways cannot belong to the same coordinate system. If you talk to your friend about meeting her at a coffee shop by walking two blocks east and four blocks south from the subway station on the 34th street and if your friend walks two blocks east and four blocks south from the subway station on the 28th street, you and your friend will end up at two different coffee shops. You can think of a coordinate system as a set of coordinates that all share the same rules or as a space in which positions are described by the same rules. There are a variety of coordinate systems that we use. Let's look at a few.

12.3.1 2D and 3D coordinate systems

A 2D coordinate system is used to specify locations in 2D space. For instance, 2D camera data mentioned earlier in this chapter is the coordinates of centers of optical markers seen from a camera described in a 2D coordinate system. It is natural to use a 2D coordinate system for the camera data since images captured by cameras are two dimensional and the data is two dimensional not three dimensional at this point. Another example is the locations of pixels. Pixels are the smallest picture elements in a digital image and their locations are described in a 2D coordinate system.

A 3D coordinate system is used to specify locations in 3D space. Users of 3D animation packages and 3D game developers are familiar with this type of coordinate systems. If we use an example from the optical mocap data generation again, 3D translational data is the 3D coordinates of markers in a 3D space described in a 3D coordinate system. 3D translational data is generated from 2D camera data using triangulation and camera positions. You will often hear the third coordinate called z-depth since characters and objects in animations could move around only in the x- and y-directions of a 2D screen space, but the z-depth allows us to see how far "into" the screen they can go. The z-depth is the essential building block of modern 3D video games and 3D animation applications.

12.3.2 Cartesian, spherical, and cylindrical coordinate systems

Cartesian coordinate systems are also called rectangular coordinate systems. With the 2D Cartesian coordinate system each location in a 2D space is specified by an ordered set of two distances: an x-coordinate and a y-coordinate. The two distances are measured from the coordinate system's origin along the x- and y-axes which are perpendicular to each other. The two coordinates are represented as (x, y). The two coordinates are "ordered" because their order is important. The x-coordinate of a point location comes first; the y-coordinate comes next. For instance, $(2, 3)$ and $(3, 2)$ specify two different locations.

With the 3D Cartesian coordinate system each location in a 3D space is specified by an ordered set of three distances from the system's origin (an x-coordinate, a y-coordinate, and a z-coordinate). The coordinates are represented as (x, y, z). The three axes of a 3D Cartesian coordinate system are orthogonal.

With the spherical coordinate system each location in a 3D space is specified by a distance and two angles. Spherical coordinate systems are used in some 3D packages to let you navigate through the world space by allowing you to change your camera position in terms of the camera's yaw and pitch and its distance from the point of interest (see Figure 12.4 for yaw and pitch). With the cylindrical coordinate system each location in a 3D space is specified by two distances and one angle.

Among the three types of 3D coordinate systems, 3D Cartesian, spherical, and cylindrical coordinate systems, we use 3D Cartesian coordinate systems most frequently.

12.3.3 Right-handed and left-handed systems

Within 3D Cartesian coordinate systems there are two kinds: right-handed 3D Cartesian coordinate systems and left-handed 3D Cartesian coordinate systems (Figure 12.1). How do we know which 3D Cartesian coordinate systems are right-handed and which ones are left-handed? If you want to test whether a 3D Cartesian coordinate system is right-handed, you want to use your right hand for the test. First, extend all of the fingers and thumb of your right hand. Align your thumb with the x-axis and your fingers with the y-axis by rotating your hand while keeping all the fingers extended. Next without rotating your hand, try to align your fingers with the z-axis by bending them. If your fingers bend in the direction that your hand would make a fist, it is a right-handed system. If you have to bend your fingers backward overextending them trying to reach the z-axis, it is not a right-handed system; it is a left-handed system.

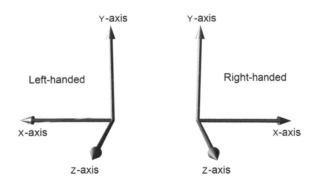

Figure 12.1 *Left-handed system and right-handed system*

Positive rotation in a 3D Cartesian coordinate system can be determined in the following way. Extend your right fingers and thumb and align the thumb with one of the x-, y-, and z-axes of a 3D Cartesian coordinate system. If it is a right-handed coordinate system, the direction of a positive rotation around the axis is the direction in which your right hand's fingers curl up to make a fist. On the other hand, in a left-handed coordinate system, the direction of a positive rotation around the axis that you selected (i.e., the axis that you align your left thumb with) is the direction in which your left hand's fingers curl up to make a fist.

Right-handed 3D Cartesian coordinate systems are used in mathematics and much more commonly used in 3D computer animation than left-handed 3D Cartesian coordinate systems.

12.3.4 Object space and world space

Each object is defined in its object space, which is also called an object coordinate system or local coordinate system. A scene containing an arbitrary number of objects is defined in the world space, which is also called the world coordinate system or global coordinate system. Object spaces are used to generate individual objects while a world space is used to build a scene with objects. Object spaces and world space work together. For instance, if an object is centered at the origin of its object space and if the object is placed at the origin of the world space (i.e., placed at (0, 0, 0) in a scene), as you expect, it will be centered at the origin of the world space. However, if an object is not centered at the origin of the object space and the object is placed at the origin of the world space, it will be off-centered in the world space.

Maya, MotionBuilder, and many other 3D animation packages use a right-handed 3D Cartesian coordinate system in which the x-axis is the horizontal axis pointing at your right, the y-axis is the vertical axis pointing upward, and the z-axis is the axis pointing toward you when you aim your camera at the origin of the coordinate system from the positive z-axis direction. However, in some other 3D packages (e.g., CAD applications) the z-axis is the vertical axis. So, when you are transferring objects from one application to another, be aware of any differences in the coordinate systems between the applications.

12.4 Order of Transformation

Order of transformation specifies in what order transformations (e.g., scaling, rotation, and translation) are applied to an object. Depending on the order of transformation, the same transformation values may yield different results.

One of the reasons why order of transformation is important is that rotation and scaling are applied to an object with respect to the origin of the object space but not the center of the object. If you scale an object that is centered at its object space's origin you will have a result that is different from scaling an identical object that has been translated away from its object space's origin prior to scaling. Similarly, if you rotate an object that is centered at its object space's origin you will have a result that is different from rotating an identical object that has been translated away from its object space's origin prior to rotation.

Some 3D applications (e.g., Maya) do not allow the user to change the order of transformation, which is inconvenient in some cases. For example, when you are making an animation of a textured ball that is rotating, stretching, and squashing while bouncing, you want to rotate the ball before you stretch or squash it as you see in Figure 12.2 (top row). However, Maya applies transformations in the fixed order – scaling, rotation, and translation. Thus scaling comes before rotation as you see the bottom row in Figure 12.2.

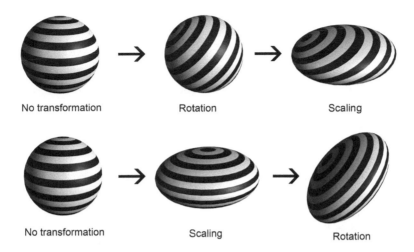

No transformation   Rotation   Scaling

No transformation   Scaling   Rotation

Figure 12.2 *Order of transformation*

To use an order of transformation of your choice which is different from Maya's fixed order a linear hierarchy of transform nodes is useful. First create three empty transform nodes (i.e., empty groups). Parent the nodes and the object so that they form a linear hierarchy in which one of the nodes is at the top, the object at the bottom, and the other two are in between. Rename the nodes appropriately (see Figure 12.3). Now use a different node for each type of transformation, instead of applying all the transformations directly to the object. Thus, do not use the scaling, rotation, and translation attributes of the object. Leave them alone. With this hierarchy of transform nodes in place, the transformations are applied from the one assigned to the lowest node in the hierarchy to the one assigned to the highest node. Make sure that the appropriate transform node is selected when you are working on transformation. For example, when you are key-framing rotation of an object, you need to have the node for rotation selected, not the node for scaling or translation nor the object itself.

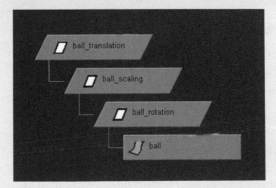

Figure 12.3 *Hierarchy of transform nodes*

When scaling an object, the order of x-, y-, and z-scaling is not important because they are independent from each other and no matter in what order x-, y-, and z-scaling is applied to an object, the result will be the same as long as the same set of x-, y-, and z-scaling values is used. Similarly regardless of the order of x-, y-, and z-translation, the result will be the same if the same set of x-, y-, and z-translation values is used. However, when rotating an object, depending on the order of x-, y-, and z-rotation, the result will be different even if the same set of x-, y-, and z-rotation values is used. We will talk more about rotation in the next two sections.

12.5 Euler Angle

Why do we need to know what Euler angles are? Because we use them. Many commercial 3D computer animation packages that are available today use Euler angles to describe and change orientations of objects. Most likely you have used Euler angles implemented in your 3D application. A set of three Euler angles and their order describe the orientation of an object in 3D space. You can think of them as the yaw, pitch, and roll of airplane navigation (Figure 12.4). Euler angles are commonly used because they are intuitive but there are some problems with Euler angles.

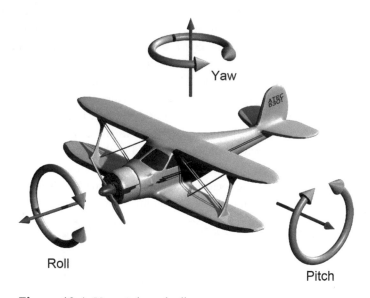

Figure 12.4 *Yaw, pitch, and roll*

One of the problems with Euler angles is that they do not represent angles uniquely. In other words, the same orientation can be represented by multiple sets of Euler angles. For instance, let's think about rotating a vertical line that is initially aligned with the y-axis. If we rotate it around the z-axis by 60 degrees first and then rotate it around the y-axis by 45 degrees, we get the same result as rotating the line around the x-axis by 60 degrees first and then rotating it around the y-axis by -45 degrees. This is because Euler angles are not unique. On the contrary any location on a globe is uniquely represented by a pair of a longitude and a latitude because two angles are enough to do the job. With Euler angles we use three rotation angles (yaw, pitch, and roll) to describe the orientation of an object. The third number seems redundant. Why do we need three rotation angles? Because we want to "roll" an object as well as yaw and pitch. When what you are rotating is a line, rolling has no effect on it. But when you are dealing with a 3D object with a volume, rolling does turn the object around. When you are piloting an airplane, you don't want to roll your airplane too much. If you rolled it by 180 degrees, you would be upside down.

As described earlier in this chapter, magnetic and mechanical motion capture systems measure joint rotation angles but optical systems do not. With an optical system the rotation angles (orientation) of each skeletal segment relative to its local coordinate system are computed from translational data at each time step. Suppose that at time t the orientation of a skeletal segment S can be represented by two different ordered sets of Euler angles, $A = (\alpha_1\beta_1\gamma_1)$ and $B = (\alpha_2\beta_2\gamma_2)$, which do not have similar values but represent identical orientations. If somehow your application decides to represent the orientation of the segment S by A at t and by angles close to B before and after t, there will be sudden changes in all or some of the rotation angles α, β, and γ at t, although in reality there is no sudden change in the orientation of the segment S at t.

The animation of the skeleton may look fine when it is played without skin. You will notice the problem when you play the animation after binding skin geometries to the skeleton or applying inverse kinematics to the skeleton. The skin geometry bound to the segment S may get twisted and the inverse kinematics applied to S may go wild at t. Such abnormalities appear as "peaks" in the graphs of the rotation channels. Delete the peaks and fill in the gaps with an appropriate method (e.g., a spline for a small gap as described in Chapter 4).

Another common problem is that an angle α can be represented as $\alpha - 360$, $\alpha + 360$, $\alpha - 720$, $\alpha + 720$, etc. For instance, -60 degrees is same as 300 degrees. Rotation angles are normally computed between 0 and 360 degrees and that can cause an issue. Suppose that your 3D character's right arm is by its side when the right shoulder's roll angle is 0 degree. You can let the arm swing by key-framing the shoulder's roll angle at -20 degrees and 20 degrees; the two degrees are interpolated for in-between frames. At the mid-point of the two key-frames, the shoulder's roll angle is 0 degree. That places the arm by the character's side. There is no problem here.

Suppose that a performer's arm swinging, similar to the arm swing of the 3D character above, is captured by an optical motion capture system. When rotation angles are computed for the skeleton's segments from translational data using angles between 0 and 360 degrees, -20 degrees would be 340 degrees. In this case, the arm of the skeleton rolls from 340 degrees to (very close to) 360 degrees and then from 0 degree to 20 degrees. Thus, over time the rotation angle increases, suddenly drops, and then increases again. The graph of the rotation angle looks discontinuous although the arm swings smoothly. Many 3D applications have a filter to clean this up. So, if you face this problem, look for a filter.

When rotation angles are computed as Euler angles in your 3D application, you may encounter gimbal lock in addition to the problems that have been discussed above. Let's look at gimbal lock in the next section.

12.6 Gimbal Lock

A *gimbal* is a mechanical device that consists of concentric rings that rotate. Gimbal rings are mounted on axes so that the axes of two adjacent rings make a perpendicular angle to each other. In aviation a gimbal measures the rotation of an aircraft using Euler angles and controls the aircraft's orientation. *Gimbal lock* occurs when two of the three gimbal rings align together and one degree of freedom is lost. Let us explain this in more detail.

Look at the gimbal in Figure 12.5. Suppose that the global space and the local space are defined in such a way that the *y*-axes point up, the *x*-axes point at your right, and the *z*-axes point toward you. The orientation of the global space does not change, whereas the local space's orientation changes as the gimbal rings rotate. Let the three gimbal rings make perpendicular angles to each other and that be the initial state of the device (Figure 12.5). Imagine rotating one ring at a time and rotating the ring back to its initial position before rotating another ring for now. The outer ring that is

Figure 12.5 *Initial state of a gimbal device*

attached to the base rotates around the *y*-axis (yaw). The middle ring rotates around the *x*-axis (pitch). The inner ring rotates around the *z*-axis (roll). Thus, the device has three degrees of freedom. Now let's suppose that you rotate the middle ring by 90 degrees so that it aligns with the outer one (Figure 12.6). Now the outer ring (yaw) and the inner ring (roll) rotate in the same manner – both rings rotate around the global *y*-axis. Thus, one degree of freedom is lost. This is gimbal lock.

Gimbal lock occurs when Euler angles are used to perform a rotation. With Euler angles, the rotation around each axis is performed one after the other. In the case of the gimbal device in Figures 10.5 and 10.6, the order of rotations is *y*, *x*, and *z*. Because of the three sequential rotations performed

Figure 12.6 *A gimbal lock*

on an object, the object's local axes may have been rotated in such a way that by the time the rotation is performed around the third axis, one axis is aligned with another. When two of an object's local axes are aligned, performing a rotation on the object creates this puzzling result which is often very annoying as well.

Users of the 3D graphics applications that rotate objects using Euler angles experience gimbal lock. Let's think about a shoulder joint's rotation using your shoulder. We will do a couple of experiments. First stretch your right arm out so that it parallels the floor. Extend your fingers and let your palm face the floor. Imagine that an airplane is pointing at the direction that your fingers are pointing at. This is your arm's initial orientation (Figure 12.7). Keep your arm stretched, that is, don't bend your arm at your elbow while performing the experiments. Moving your arm up and down by rotating it at your shoulder is pitch (x-axis rotation). Moving your arm back and forth by rotating it at your shoulder is yaw (y-axis rotation). Twisting your arm is roll (z-axis rotation). Let's use a right-handed 3D Cartesian space. Recall that in a right-handed space if you align your right thumb with one of the axes, pointing the thumb at the positive axis direction, and curl your fingers up, then the direction that an object rotates by a given positive rotation angle is the direction that your curled fingers point to.

Apply a yaw of 90 degrees to your arm, that is, move your arm to your front keeping your palm facing the floor (Figure 12.8a). Then apply a pitch of 90 degrees. Thus twist your arm so that your palm faces your left while keeping your arm stretched in front of you (Figure 12.8b). And apply a roll of −90 degrees. That brings your arm by your side (Figure 12.8c). Varying the roll angle, say between −70 and −110 degrees, lets you swing your arm back and forth. The first experiment has been completed. Remember three steps (a 90 degree yaw, a 90 degree pitch, and a −90 degree roll) brought your arm by your side and you were able to swing your arm. Let's try another experiment.

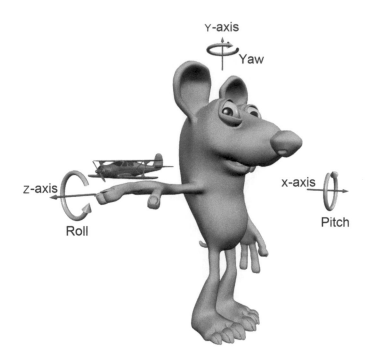

Figure 12.7 *Initial state*

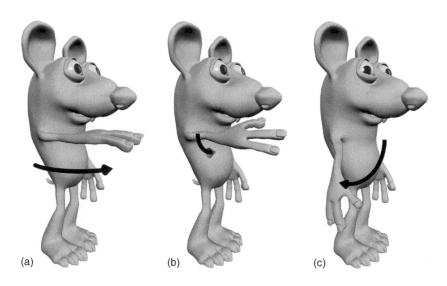

Figure 12.8 *(a) 90 degree yaw; (b) 90 degree pitch; and (c) −90 degree roll*

Bring your right arm back to the initial orientation (Figure 12.7). This time apply a pitch of 90 degrees. That moves your arm down (i.e., places your arm by your side) (Figure 12.9a). Recall that in the first experiment it took three steps, a yaw, a pitch, and a roll, to bring your arm by your side. This time a single step, a pitch, brought your arm to the same position. This is a much more intuitive way to place a character's arm by its side but in this way you face a problem, gimbal lock. Recall what happened to the gimbal device when the outer and middle rings aligned (Figure 12.6). By placing your arm by your side using a pitch of 90 degrees, you let the local y-axis align with the local z-axis (Figure 12.9a). As the result both yaw and roll have become aligned as you were twisting your arm (Figure 12.9b). Because one degree of freedom is lost there is no rotation that allows you to swing your arm as you did in the first experiment. This can happen to the arms of a 3D character if your application is implemented with Euler angles.

Gimbal lock does not occur in the application that uses quaternions, described in the next section, instead of Euler angles. When you experience gimbal lock, check if your 3D application has any tools that prevent gimbal lock. For instance, Maya has an option to use quaternion rotation, instead of Euler angle rotation.

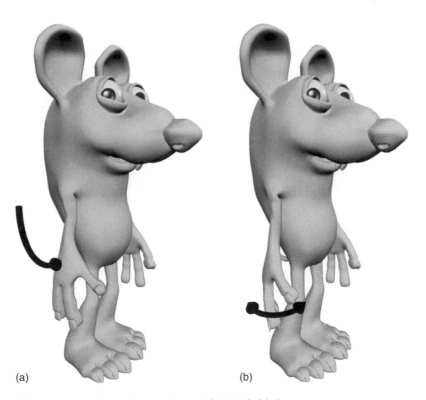

(a)   (b)

Figure 12.9 *(a) 90 degree pitch; and (b) Gimbal lock*

12.7 Quaternions

Euler angles can be understood as pitch, yaw, and roll by imagining being a pilot of an airplane. Quaternions are not so easily understood as Euler angles because quaternions are four dimensional. We are residents of a 3D world and it's rather hard for us to imagine 4D things. But let us try to give you some ideas about what quaternions are. First of all, quaternions are an extension of complex numbers. A complex number is the number of the form:

$a + bi,$

where a and b are real numbers and i is the imaginary unit.

The imaginary unit i has a peculiar nature (i.e., $i^2 = -1$). Thus, $2i \times 3i = -6$. Quaternions are a little bit more complex than complex numbers. They have the form of:

$a + bi + cj + dk,$

where a, b, c and d are real numbers and $i^2 = j^2 = k^2 = -1$.

Secondly quaternion multiplications are not commutative. In usual algebra of real numbers that we are accustomed to, multiplications are commutative (i.e., $xy = yx$). Thus, 2×3 equals to 3×2. However, $a \times b$ does not equal $b \times a$ when a and b are quaternions.

Most importantly, quaternions do not suffer the drawbacks that Euler angles do; therefore, they provide us with a wonderful method to compute 3D rotations. Quaternions cannot be easily visualized or understood because they are four dimensional. However, implementing rotations (and interpolation of rotations) with quaternions and conversions between quaternions and Euler angles are not so difficult. Therefore more and more 3D software packages are offering implementations of quaternions that supplement or replace the Euler angle representation.

Bibliography

Alexander, R. McNeill. *Exploring Biomechanics: Animals in Motion*. Scientific American Library, New York, NY, 1992.

Calais-Germain, Blandine. *Anatomy of Movement*. Eastland Press, Vista, CA, 1993.

Caputp, Tony C. *Visual Storytelling: The Art and Technique*. Watson-Guptill Publications, New York, NY, 2003.

Edgerton, Harold E and James R. Killan, Jr. *Moments of Vision*. The MIT Press, Cambridge, MA, 1979.

Hass, Robert Bartlett. *Muybridge: Man in Motion*. University of California Press, Berkeley, CA, 1976.

Hendricks, Gordon. *Eadweard Muybridge: The Father of the Motion Picture*. Grossman Publishers, New York, NY, 1975.

Heraldson, Donald. *Creators of Life: A History of Animation*. Drake Publishers, Inc., New York, NY, 1975.

Jussim, Estelle and Gus Karafas. *Stopping Time: The Photographs of Harold Edgerton*. Harry N. Abrams, New York, NY, 1987.

Kardong, Kenneth V. *Vertebrates: Comparative Anatomy, Function, Evolution*. McGraw Hill, New York, NY, 2002.

Maltin, Leonard. *Of Mice and Magic: A History of American Animated Cartoons*. Revised Edition. Plume, New York, NY, 1987.

Menache, Alberto. *Understanding Motion Capture for Computer Animation and Video Games*. Morgan Kaufmann, San Francisco, CA, 1995.

Muybridge, Eadweard. *Animals in Motion*. Dover Publications, Inc., New York, NY, 1957.

Muybridge, Eadweard. *The Human Figure in Motion*. Dover Publications, Inc., New York, NY, 1955.

Parent, Rick. *Computer Animation: Algorithms and Techniques*. Morgan Kaufmann Publishers, San Francisco, CA, 2002.

Sieg, Kay W. *Illustrated Essentials of Musculoskeletal Anatomy*. Megabooks, Inc., Gainesville, FL, 2002.

Solomon, Charles. *Enchanted Drawings: History of Animation*. Alfred A. Knopf, New York, NY, 1989.

Thomas, Frank and Ollie Johnston, *Disney Animation: The Illusion of Life*. Abbeville Press, New York, NY, 1981.

Vogel, Steven. *Life's Devices*. Princeton University Press, Princeton, NJ, 1988.

Appendix A: Shot List for Juggling Cow

Scene	Shot	Shot name	Description
Scene1			
	Shot 1	JC0101	Cow enters the back of an old barn
	Shot 2	JC0102	Still shot of bowling ball, chainsaw and pool cue.
	Shot 3	JC0103	Cow jumps as if startled blend
	Shot 4	JC0104	Cow walks toward objects
Scene 2			
	Shot 1	JC0201	Cow timidly picks up the bowling ball
	Shot 2	JC0202	Cow throws bowling ball in the air a few times, gaining confidence
	Shot 3	JC0203	Cow uses hoof to toss other objects into the air
Scene 3			
	Shot 1	JC0301	Cow is juggling all three objects
	Shot 2	JC0302	See cow from over head and hear a loud off screen yell of "Bessy! Milkin' Time!"
	Shot 3	JC0303	Cow is startled and drops all the props
	Shot 4	JC0304	Cow soberly walks back out of the barn

Appendix B: Sample Mocap Production Pipeline and Data Flow Chart

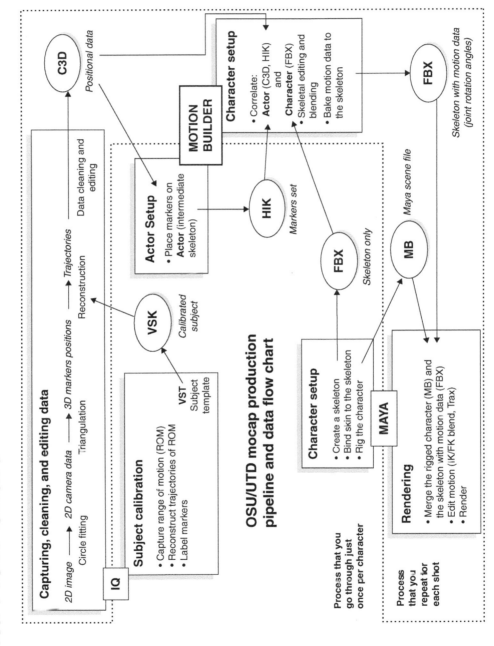

OSU/UTD mocap production pipeline and data flow chart

Capturing, cleaning, and editing data

2D image → 2D camera data → 3D markers positions → Trajectories

Circle fitting   Triangulation   Reconstruction   Data cleaning and editing

Subject calibration
- Capture range of motion (ROM)
- Reconstruct trajectories of ROM
- Label markers

VST Subject template

IQ

VSK — *Calibrated subject*

C3D — *Positional data*

Actor Setup
- Place markers on **Actor** (intermediate skeleton)

HIK — *Markers set*

MOTION BUILDER

Character setup
- Correlate:
  Actor (C3D, HIK) and **Character** (FBX)
- Skeletal editing and blending
- Bake motion data to the skeleton

FBX — *Skeleton with motion data (joint rotation angles)*

FBX — *Skeleton only*

Character setup
- Create a skeleton
- Bind skin to the skeleton
- Rig the character

MAYA

MB — *Maya scene file*

Rendering
- Merge the rigged character (MB) and the skeleton with motion data (FBX)
- Edit motion (IK/FK blend, Trax)
- Render

Process that you go through just once per character

Process that you repeat for each shot

Glossary

2D coordinate system   A 2D coordinate system is used to specify locations in 2D space.

3D Cartesian coordinate system   With a 3D Cartesian coordinate system each location in a 3D space is specified by an ordered set of three distances from the system's origin (an x-coordinate, a y-coordinate, and a z-coordinate). The coordinates are represented as (x, y, z). The three axes of a 3D Cartesian coordinate system are orthogonal.

3D coordinate system   A 3D coordinate system is used to specify locations in 3D space.

Animatic   A series of drawings and/or rough animations with sound which are created to test timing and continuity.

Bake   To permanently assign key-frames to a keyable attribute of an object that is driven by a method other than key-framing.

Broken hierarchy   A hierarchy that consists of disjointed chains of joints. An example is a skeleton where the arms and legs are not attached to the hips and spine, but can still follow them if needed.

Calibrate   To determine or check the accuracy of a system.

Coordinates   Coordinates are an "ordered" set of values which specify a location relative to an origin.

Coordinate system   A coordinate system is a system that defines how coordinates are assigned to each point in the space.

Creep   Slow, small movements.

Cylindrical coordinate system   In a cylindrical coordinate system each location in a 3D space is specified by two distances and one angle.

Degrees of freedom (DOF)   DOF of an object is the number of independent variables that specifies the location of the object. In 3D space, a rigid object can have the maximum of six DOF that are the x-, y-, and z-translations and x-, y-, and z-rotations.

Ease in/ease out   See slow in and slow out.

Effector (end effector)   An effector is the end point in a chain of joints that is placed at a desired position. Inverse kinematics is used to compute the rotation angles of the middle joints in the chain to reach the position. (See Inverse kinematics.)

Euler angles   Euler angles were developed by an 18th-century mathematician, Leonhard Euler. The orientation of an object in 3D space is specified by a sequence of three rotations described by the Euler angles.

.fbx   A 3D data format designed to describe animation scenes and is supported by many 3D animation software packages to transfer files among them.

Floating motion   Motion that looks unnaturally fluid because the accelerations and velocities of the motion have been compromised by data editing or filtering.

Forward kinematics   Forward kinematics is a method of animating a skeleton where the animator specifies and keys the position of every joint in the skeleton in one frame, moves to another frame, and repeats it until the desired motion is achieved.

Gimbal lock   Gimbal lock occurs when two of the three axes of a 3D Cartesian system align together and one degree of freedom is lost.

Hierarchy   A system of relationships among elements where each element is a subordinate (a child) of a single dominant element (a parent). The element at the top of a hierarchy is called the root. Each element (except for the root) has one parent and an arbitrary number of child elements. A transformation applied to a parent is applied to its child as well, but a transformation applied to a child is not applied to the parent. Geometries, markers, and joints are often structured in hierarchies.

Inbetweens   Inbetweens fill the gaps between key-frames. In the production of a traditional hand-drawn animation, inbetweens are drawn by less experienced animators while key-frames are drawn by skilled animators. In the production of a 3D animation the parameter values for inbetweens are generated by interpolating the parameter values of the key-frames.

Inverse kinematics (IK)   Inverse kinematics is a method of animating a skeleton where the animator specifies only the position of the end effector. The software calculates all the rotation angles of the middle joints in the chain to reach the position of the end effector.

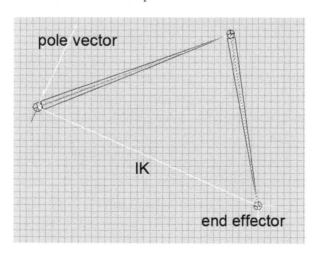

Key See Key-frame.

Key-frame A key is a defining moment of a motion. In a 3D animation sequence a key is created where an attribute (a parameter value) of an entity (e.g., an object, a light, a camera) is at its extreme. A key for an entity's attribute is specified by two values: the time of the key moment (frame number) and the parameter value at the key moment. The number of keys in an animation sequence depends on how complicated the movement is.

Left-handed system If you can align your left thumb with the x-axis of a 3D Cartesian coordinate system, your left index finger with the y-axis, and your left middle finger with the z-axis, then the coordinate system is left-handed.

Marker See Sensors.

Mocap Short form of motion capture.

Object space Each object is defined in its object space, which is also called an object coordinate system or local coordinate system.

Occlusion Hiding of one object. In the case of optical motion capture, hiding or covering a marker from the view of one or several cameras.

Order of transformation Order of transformation specifies in what order transformations (e.g., scaling, rotation, and translation) are applied to an object. Depending on the order of transformation, the same transformation values may yield different results.

Pole vector A vector that is used with an IK to define the plane that the middle joints in the IK chain lie on. (See Inverse kinematics.)

Principles of animation While animation was maturing from a novelty to an art form enjoyed by many families in the 1930s, animators at the Disney Studios created 12 principles of animation. The principles were developed to guide production of traditional hand-drawn animation, especially character animation, and to train younger animators. These principles can help us create believable characters and situations for key-frame animation and animation driven by mocap data. The principles are: (1) Squash and stretch, (2) Anticipation, (3) Staging, (4) Straight ahead action and pose to pose, (5) Follow through and overlapping action, (6) Slow in and slow out, (7) Arcs, (8) Secondary action, (9) Timing, (10) Exaggeration, (11) Solid drawings, and (12) Appeal. Read *Disney Animation: The Illusion of Life* by Frank Thomas and Ollie Johnston to learn more about the principles.

Quaternion Mathematically speaking quaternions are a non-commutative extension of complex numbers. Quaternions are used to compute angles and rotations of objects in 3D space in place of Euler angles because quaternions do not suffer from gimbal lock.

Render To create an image from descriptions of 3D objects in a scene.

Right-handed system If you can align your right thumb with the x-axis of a 3D Cartesian coordinate system, your right index finger with the y-axis, and your right middle finger with the z-axis, then the coordinate system is right-handed.

Root joint The highest joint in a skeleton's hierarchy. A skeleton can have only one root joint, while it can have multiple chains of joints that branch out.

Rotoscoping Rotoscoping is a method of producing an animation by drawing, frame by frame, over live action reference.

Script A narrative structure that formats incidents and dialog for film making.

Sensors A type of device that transmits some form of information or from which information can be derived. Magnetic sensors give both positions and orientations while optical markers give positions only.

Shot list A list of actions or motions that will be compiled together to create a scene.

Skeleton A hierarchically articulated structure of joints. It is used for posing and animating deformable objects (skin geometries) that are bound to the structure.

Slow in and slow out One of the principles of animation. Rather than instantly moving at full speed and just as suddenly stopping at the close of the action, apply slowly in at its start and slow out at its closure builds in acceleration and deceleration into the action, which creates more realism in the motions of animated characters and objects. In 3D animation slow in and slow out are often achieved by use of the spline interpolation. With a properly shaped spline curve, a spline interpolation method, instead of a linear interpolation method, is applied to data in order to generate inbetweens from key-frames.

Spherical coordinate system In a spherical coordinate system each location in a 3D space is specified by a distance and two angles.

Squash and stretch One of the principles of animation. It gives a non-rigid object an organic flexibility by distorting its shape in accordance to the stressor acting upon it or the degree of physicality of its action.

Storyboard A set of drawings and accompanying dialog, which serves as a 2D visual representation of the script.

Thumbnail sketch   A quick, small sketch that's used to illustrate basic visual ideas, such as the actions of the characters and camera positions.

T-pose   A generic standing pose with feet shoulder width apart, back straight, and arms out to the side with palms down. It is usually a starting position for human motion capture performers.

World space   A scene containing an arbitrary number of objects is defined in the world space, which is also called the world coordinate system or global coordinate system.

Index

T - #0614 - 071024 - C0 - 246/189/10 - PB - 9780240810003 - Matt Lamination